Advances in Intelligent Systems and Computing

Volume 699

Series editor

Janusz Kacprzyk, Polish Academy of Sciences, Warsaw, Poland
e-mail: kacprzyk@ibspan.waw.pl

The series "Advances in Intelligent Systems and Computing" contains publications on theory, applications, and design methods of Intelligent Systems and Intelligent Computing. Virtually all disciplines such as engineering, natural sciences, computer and information science, ICT, economics, business, e-commerce, environment, healthcare, life science are covered. The list of topics spans all the areas of modern intelligent systems and computing.

The publications within "Advances in Intelligent Systems and Computing" are primarily textbooks and proceedings of important conferences, symposia and congresses. They cover significant recent developments in the field, both of a foundational and applicable character. An important characteristic feature of the series is the short publication time and world-wide distribution. This permits a rapid and broad dissemination of research results.

More information about this series at http://www.springer.com/series/11156

Peeyush Chandra · Debasis Giri
Fagen Li · Samarjit Kar · Dipak Kumar Jana
Editors

Information Technology and Applied Mathematics

ICITAM 2017

Springer

Editors
Peeyush Chandra
Department of Mathematics and Statistics
Indian Institute of Technology Kanpur
Kanpur, Uttar Pradesh
India

Samarjit Kar
Department of Mathematics
National Institute of Technology Durgapur
Durgapur, West Bengal
India

Debasis Giri
Department of Computer Science and Engineering
Haldia Institute of Technology
Haldia, West Bengal
India

Dipak Kumar Jana
Department of Applied Sciences
Haldia Institute of Technology
Haldia, West Bengal
India

Fagen Li
School of Computer Science and Engineering
University of Electronic Science and Technology of China
Chengdu
China

ISSN 2194-5357 ISSN 2194-5365 (electronic)
Advances in Intelligent Systems and Computing
ISBN 978-981-10-7589-6 ISBN 978-981-10-7590-2 (eBook)
https://doi.org/10.1007/978-981-10-7590-2

Library of Congress Control Number: 2018936624

Printed on acid-free paper

This Springer imprint is published by the registered company Springer Nature Singapore Pte Ltd.
part of Springer Nature
The registered company address is: 152 Beach Road, #21-01/04 Gateway East, Singapore 189721, Singapore

Preface

International Conference on Information Technology and Applied Mathematics (ICITAM 2017) was held at the Haldia Institute of Technology, Haldia, from October 30 to November 01, 2017. Haldia is a city and a municipality in Purba Medinipur in the Indian state of West Bengal, India, and Haldia Institute of Technology is a premier institution training engineers and computer scientists for the past several years. It has gained its reputation through its institutional dedication to teaching and research.

In response to the call for papers for ICITAM 2017, 65 papers were submitted for presentation and inclusion in the proceedings of the conference. The papers were evaluated and ranked on the basis of their significance, novelty, and technical quality by at least two reviewers per paper. After a careful blind refereeing process, 16 papers were selected for inclusion in the conference proceedings. The papers cover current researches in Intelligent Systems, Soft Computing, Machine Learning, Natural Language Processing, Image and Video Processing, Computer Network and Security, Cryptography, and Data Hiding, Rough Set, Fuzzy Logic, Operations Research, Optimization, Uncertain Theory and Applications, etc. ICITAM 2017 had eminent personalities from both India and abroad (China and Turkey) who will deliver invited talks. The speakers from India are recognized leaders in government, industry, and academic institutions like Indian Statistical Institute, Kolkata; IIT Kharagpur; Central University of Rajasthan; IIT Kanpur; Harish-Chandra Research Institute; Vidyasagar University. All of them are involved in research dealing with the current issues of interest related to the theme of the conference. The conference included one keynote talk by Prof. Sankar Kumar Pal (Indian Statistical Institute, Kolkata, India) and nine invited talks by Prof. Arun K. Pujari (Central University of Rajasthan, India), Prof. P. D. Srivastava (IIT Kharagpur, India), Prof. Baoding Liu (Tsinghua University, China), Prof. A. Okay Celebi (Yeditepe University Istanbul, Turkey), Prof. Kalyan Chakraborty (Harish-Chandra Research Institute, India), Prof. Manoranjan Maiti (Vidyasagar University, India), and Dr. Nilanjan Dey (Techno India College of Technology, India).

A conference of this kind would not be possible to organize without the full support from different people across different committees. All logistics and general organizational aspects were looked after by the organizing committee members who spent their time and energy in making the conference a reality. We also thank all the technical program committee members and external reviewers for thoroughly reviewing the papers submitted for the conference and sending their constructive suggestions within the deadlines. Our hearty thanks go to Springer for agreeing to publish the proceedings in its Advances in Intelligent System and Computing series.

We are indebted to Defense Research and Development Organization (DRDO) and Haldia Institute of Technology, Haldia, India, for sponsoring the event. Their support has significantly helped in raising the profile of the conference.

Last but not least, our sincere thanks go to all authors who submitted papers to ICITAM 2017 and to all speakers and participants. We sincerely hope that the readers will find the proceedings stimulating and inspiring.

Kanpur, India — Peeyush Chandra
Haldia, India — Debasis Giri
Chengdu, China — Fagen Li
Durgapur, India — Samarjit Kar
Haldia, India — Dipak Kumar Jana

Committee

Patron

Lakshman Seth, Chairman, Haldia Institute of Technology, Haldia, India

General Co-chairs

P. D. Srivastava, IIT Kharagpur, India
Arun K. Pujari, Central University of Rajasthan, India

Program Co-chairs

Peeyush Chandra, IIT Kanpur, India
Debasis Giri, Haldia Institute of Technology, Haldia, India
Fagen Li, University of Electronic Science and Technology of China, China
Samarjit Kar, National Institute of Technology Durgapur, India
Dipak Kumar Jana, Haldia Institute of Technology, Haldia, India

Organizing Committee

Asish Lahiri, Haldia Institute of Technology, India
Sayantan Seth, Haldia Institute of Technology, India
M. N. Bandyopadhyay, Haldia Institute of Technology, India
Asit Kumar Saha, Haldia Institute of Technology, India
Anjan Mishra, Haldia Institute of Technology, India
Sudipta Kumar Basu, Haldia Institute of Technology, India
Debasis Giri, Haldia Institute of Technology, India
Debasis Das, Haldia Institute of Technology, India
Dipak Kumar Jana, Haldia Institute of Technology, India
Palash Ray, Haldia Institute of Technology, India
Mihir Baran Bera, Haldia Institute of Technology, India
Pronab Ghoswami, Haldia Institute of Technology, India

Anupam De, Haldia Institute of Technology, India
Snehasish Kumar Karan, Haldia Institute of Technology, India

Organizing Chair

Subhankar Joardar, Haldia Institute of Technology, India
Sourav Mandal, Haldia Institute of Technology, India

Organizing Co-chairs

Joyeeta Majumdar, Haldia Institute of Technology, India
Sk. Arif Ahmed, Haldia Institute of Technology, India

Publication Committee

Anupam De, Haldia Institute of Technology, India
Sk. Arif Ahmed, Haldia Institute of Technology, India
Nilanjan Dey, Techno India College of Technology, Kolkata, India
Samarjit Kar, National Institute of Technology Durgapur, India
Debasis Giri, Haldia Institute of Technology, India

Technical Program Committee

Dipanwita Roy Chowdhury, Indian Institute of Technology Kharagpur, India
Oscar Castillo, Tijuana Institute of Technology, Mexico
Heinrich Begehr, Freie Universitat Berlin, Berlin, Germany
Gennadii V. Demidenko, Sobolev Institute of Mathematics, Russia
Nima Kazemi, University of Malaya (UM), Malaysia
Sini Guo, Beijing University of Chemical Technology, China
Hao Hu, Beijing University of Chemical Technology, China
Samarjit Kar, National Institute of Technology Durgapur, Durgapur, India
Sk. Arif Ahmed, Haldia Institute of Technology, Haldia, India
P. D. Srivastava, Indian Institute of Technology Kharagpur, Kharagpur, India
Sourav Mandal, Haldia Institute of Technology, Haldia, India
Chayan Halder, Barasat State University, Kolkata, India
Dipak Kr. Jana, Haldia Institute of Technology, Haldia, India
Pradip Kundu, National Institute of Technology Durgapur, India
Sachin Shaw, Botswana International University of Science and Technology, Botswana
Soumen Saha, University of Calcutta, Kolkata, India
Debasis Giri, Haldia Institute of Technology, Haldia, India
Tanmay Maitra, KIIT, Bhubaneswar, Orissa, India
Biswapati Jana, Vidyasagar University, Midnapore, India
Alip Mohammed, The Petroleum Institute, Abu Dhabi
Nilanjan Dey, Techno India College of Technology, Kolkata, India
Biswajit Mondal, Raja N. L. Khan Women's College, Midnapore, India
Hongguang Ma, Beijing University of Chemical Technology, China
Pravati Swain, National Institute of Technology Goa, India
Subhabrata Barman, Haldia Institute of Technology, Haldia, India

Message from the General Chairs

As we all are aware, mathematics has always been a discipline of interest not only to theoreticians but also to all practitioners irrespective of their specific profession. Be it science, technology, economics, commerce, or even sociology, new mathematical principles and models have been emerging and helping in new research and in drawing inferences from practical data as well as through logic. Past few decades have seen enormous growth in applications of mathematics in different areas which are multidisciplinary in nature. Cryptography, security, and signal processing are such areas which have got more focus recently due to the need of securing communication while connecting with others. With emerging computing facilities and speeds, a phenomenal growth has happened in problem-solving area. Earlier, some observations were made and conjectures were drawn which remained conjectures till somebody either could prove it theoretically or found counterexamples. But today, we can write algorithms and use computers for long calculations, verifications, or generation of huge amount of data. With available computing capabilities, we can find factors of very large integers of the size of hundreds of digits; we can find inverses of very large size matrices and solve a large set of linear equations and so on. Thus, mathematics and computations have become more integrated areas of research these days, and it was thought to organize an event where thoughts may be shared by researchers and new challenging problems could be deliberated for solving these.

Apart from many other interdisciplinary areas of research, cryptography and security have emerged as one of the most important areas of research with discrete mathematics as a base. Several research groups are actively pursuing research on different aspects of cryptology not only in terms of new crypto-primitives and algorithms but a whole lot of concepts related to authentication, integrity, and security proofs/protocols are being developed, many times with open and competitive evaluation mechanism to evolve standards.

As conferences, seminars, and workshops are the mechanisms to share knowledge and new research results giving us a chance to get new innovative ideas for futuristic needs as threats and computational capabilities of adversaries are ever increasing, it was thought appropriate to organize the present conference focused on mathematics and computations covering theoretical as well as practical aspects of research, cryptography and security being one of these.

Eminent personalities working in mathematical sciences and related areas have been invited from abroad as well as from within the country, who would deliver invited talks and tutorials for participants. The talks by these speakers intend to cover a wide spectrum, viz intelligent computing, uncertainty, fuzzy. The conference is spread over 3 days (October 30–November 01, 2017). The main conference is planned with special talks by experts and paper presentations in each session.

I hope that the conference would meet the aspirations of the participants and meet its objective of ideas and current research being shared and new targets/problems identified; more so the young researchers and students would get new directions to pursue their future research.

P. D. Srivastava
IIT Kharagpur, India

Arun K. Pujari
Central University of Rajasthan, India

Message from the Program Chairs

It is a great pleasure for us to organize the International Conference on Information Technology and Applied Mathematics (ICITAM 2017) held from October 30 to November 01, 2017, at the Haldia Institute of Technology, Purba Medinipur, West Bengal, India. Our main goal is to provide an opportunity to the participants to learn about contemporary research in applied mathematics and computing, and exchange ideas among themselves and with experts present in the conference as the plenary as well as invited speakers. With this aim in mind, we carefully selected the invited speakers. It is our sincere hope that the conference will help the participants in their research and training and open new avenues for work for those who are either starting their research or looking for extending their area of research to a different area of current research in applied mathematics and computing.

The conference included one keynote talk by Prof. Sankar Kumar Pal (Indian Statistical Institute, Kolkata, India) and nine invited talks by Prof. Arun K. Pujari (Central University of Rajasthan, India), Prof. P. D. Srivastava (IIT Kharagpur, India), Prof. Baoding Liu (Tsinghua University, China), Prof. A. Okay Celebi (Yeditepe University Istanbul, Turkey), Prof. Kalyan Chakraborty (Harish-Chandra Research Institute, India), Prof. Manoranjan Maiti (Vidyasagar University, India), and Dr. Nilanjan Dey (Techno India College of Technology, India).

After an initial call for papers, 65 papers were submitted for presentation at the conference. All submitted papers were sent to external referees, and after refereeing, 22 papers were recommended for publication for the conference proceedings that will be published by Springer in its Advances in Intelligent System and Computing series.

We are grateful to the speakers, participants, referees, organizers, sponsors, and funding agencies for their support and help without which it would have been impossible to organize the conference, the workshops, and the tutorials. We owe our gratitude to the volunteers who work behind the scenes tirelessly in taking care of the details in making this conference a success.

Peeyush Chandra
Debasis Giri
Fagen Li
Samarjit Kar
Dipak Kumar Jana

Contents

About the Editors

Prof. Peeyush Chandra is a Retired Professor at the Department of Mathematics and Statistics, Indian Institute of Technology (IIT) Kanpur, India, and Past President of the Indian Mathematical Society and Indian Society of Theoretical and Applied Mechanics. He completed his M.Sc. and Ph.D. at IIT Kanpur, India, and B.Sc. at Meerut College, Meerut University, India. His research interests include bio-fluid mechanics, fluid mechanics (low Reynolds number flow) lubrication theory, magnetic fluids, ecological modeling, and mathematical epidemiology. He has published over 100 articles in refereed journals and proceedings of leading national and international conferences. He has over 35 years of teaching and over 45 years of research experience.

Dr. Debasis Giri is Dean of the School of Electronics, Computer Science, and Informatics and a Professor at the Department of Computer Science and Engineering, Haldia Institute of Technology, India. He completed his M.Tech. and M.Sc. at IIT Kharagpur, India, and also his doctorate at IIT Kharagpur, India. He was ranked tenth in the country (India) in the Graduate Aptitude Test in Engineering in 1999. He has published over 50 papers in international journals/conferences. His current research interests include cryptography, information security, e-commerce security, and design and analysis of algorithms. He is an editorial board member and reviewer for numerous international journals. He is also a program committee member for various international conferences. He is a life member of the Cryptology Research Society of India and the International Society for Analysis, its Applications and Computation (ISAAC).

Dr. Fagen Li is a Professor at the School of Computer Science and Engineering, University of Electronic Science and Technology of China (UESTC), Chengdu, P.R. China. He received his Ph.D. in Cryptography from Xidian University, Xi'an, in 2007, and from 2008 to 2009, he was a Postdoctoral Fellow at the Future University Hakodate, Hokkaido, Japan, which is supported by the Japan Society for the Promotion of Science (JSPS). He worked as a Research Fellow at the Institute of Mathematics for Industry, Kyushu University, Fukuoka, Japan, from 2010 to

2012. His recent research interests include cryptography and network security. He has published more than 80 papers in international journals and conferences. He is a member of IEEE.

Dr. Samarjit Kar is associated with the Department of Mathematics, National Institute of Technology Durgapur, India. He received his Ph.D. from Vidyasagar University, Midnapore, India, and his M.Sc. and B.Sc. from Calcutta University, Kolkata, India. His research interests include operations research and optimization, multicriteria decision-making, financial modeling, soft computing, and uncertainty modeling. He has published 74 journal papers, 2 books, and 5 edited volumes. He has supervised 12 Ph.D. students to date.

Dr. Dipak Kumar Jana is an Associate Professor at the Department of Applied Science, Haldia Institute of Technology, India. He received his Ph.D. (Science) from IIEST, Shibpur, and his M.Sc. and B.Sc. from Vidyasagar University, India. He has 12 years of teaching and 5 years of research experience. His areas of interest include operations research, inventory, uncertainty theory, bio-mathematics, transportation, rough sets, optimal control, and fuzzy inference. He has published 6 books, 2 chapters, 37 journal papers, and 16 conference papers.

Part I
Information Technology

Identity-Based Combined Public Key Schemes for Signature, Encryption, and Signcryption

Yuyang Zhou, Zhaoqi Li, Fei Hu and Fagen Li

Abstract Signature, encryption, and signcryption are three basic cryptographic primitives in the public key cryptography. In this paper, we discuss identity-based combined public key schemes in three cryptographic primitives environment, signature, encryption, and signcryption. The advantage of using combined public key scheme is to reduce the task of key management, where the same key pair is applied for signature, encryption, and signcryption. We give an identity-based combined signature and encryption (IBCSE) method based on Cha and Cheon's signature and Boneh and Franklin's encryption. In addition, we point out that the security notions for combined signature, encryption and signcryption defined by Paterson et al. in ASIACRYPT 2011 are too strong. We define relatively weak but more reasonable security notions for identity-based combined signature, encryption and signcryption (IBCSESC). We give a weakly secure IBCSESC scheme that satisfies our weak security notions and a strongly secure IBCSESC scheme that satisfies strong security notions.

Keywords Identity-based cryptography · Combined public key scheme
Signature · Encryption · Signcryption

1 Introduction

Confidentiality, integrity, authentication, and non-repudiation are four important security properties in the cryptography. Confidentiality guarantees messages secret from others than those who are allowed to see it. Integrity keeps that the messages have not been modified by unauthorized persons. Authentication guarantees that the communicating party is the correct one that it declares to be. Non-repudiation assures that a communicating party cannot deny previous commitments or actions. Signature, encryption, and signcryption are three basic cryptographic primitives in the cryptog-

Y. Zhou · Z. Li · F. Hu · F. Li (✉)
University of Electronic Science and Technology of China, Chengdu 611731, China
e-mail: fagenli@uestc.edu.cn

P. Chandra et al. (eds.), *Information Technology and Applied Mathematics*,
Advances in Intelligent Systems and Computing 699,
https://doi.org/10.1007/978-981-10-7590-2_1

raphy. Signature is used to obtain the integrity, authentication, and non-repudiation. Encryption is used to obtain the confidentiality. Signcryption [1] is used to simultaneously obtain confidentiality, integrity, authentication, and non-repudiation by one operation. Of course, we also can use the "encrypt-then-sign" (EtS) or "sign-then-encrypt" (StE) methods to obtain the signcryption function. The advantage of signcryption is that it gets a lower cost than that required by the EtS and StE methods. The formal security model for signcryption was studied by An et al. [2] and Baek et al. [3], respectively.

Identity-based cryptography (IBC) is introduced by Shamir [4] in 1984. The IBC uses any binary string, such as an email address and an IP address, as a user's public key. There is a trusted third party named private-key generator (PKG) who is in charge of generating a private key for the user. The IBC eliminates the public key certificates and removes the attached management problems, such as generation, distribution, storage, verification, and revocation. On the other hand, the IBC has the key escrow problem since the PKG knows all users' private keys. Several efficient identity-based signatures (IBSs) [5–7], identity-based encryption (IBE) [8, 9], and identity-based signcryption (IBSC) [10–12] have been proposed.

The formal method of combined public key schemes was started by Haber and Pinkas [13] in 2001. They discussed if it is secure to apply the same key pair for two (or more) different cryptographic schemes. For example, it is dangerous to apply the same key pair in textbook RSA signing and decryption. Haber and Pinkas showed that in many cases, the key reuse for two cryptographic schemes, e.g., a signature scheme and an encryption scheme, does not compromise their security. For a signature, they proved that an adversary who has access to a decryption oracle of an encryption using a related private key does not weaken the security strength of the signature. Analogously, for an encryption, they proved that an adversary who has access to a signature oracle of a signature using a related private key does not weaken the security strength of the encryption. The merit of key reuse is that it can reduce the work of key management. For example, if a user is a signer, a decrypter, a signcrypter, and an unsigncrypter, it needs four different key pairs. It is a heavy task to manage these keys, including key storage, key verification, key update, and key revocation. It will become more convenient if we apply the same key pair for all signature, encryption, and signcryption operations. The global standard EMV [14] also permits the same key pair to be reused in signature and encryption operations. Therefore, it is significant and important for us to study if the same key pair can be applied in two (or more) different cryptographic primitives. In ACISP 2011, Fan et al. [15] proved that a user can apply the same key pair in both signcryption and unsigncryption operations in Zheng's signcryption [1]. They also proved that Zheng's signcryption permits a user to securely signcrypt a plaintext to itself. In ASIACRYPT 2011, Paterson et al. [16] construct a combined signature, encryption and signcryption (CSESC) scheme using the "sign-then-tag-based encrypt" (StTE) method. This is first time to discuss key reuse problem in three cryptographic schemes. Paterson et al. [16] also defined a very strong security notion for a CSESC scheme that has the following characteristics: (1) An adversary is permitted to access a signcryption oracle in an unrestricted manner to break the unforgeability of signature component of a CSESC scheme; (2) the adversary also can have access to an unsigncryption oracle in an unrestricted manner to break the confidentiality of encryption component of a CSESC scheme; (3) the adversary can make a signature oracle in an unrestricted manner to break the unforgeability of signcryption component of a CSESC scheme; and (4) the adversary can make a decryption oracle in an unrestricted manner to break the confidentiality of signcryption component of a CSESC scheme. We think such definition is too strong since that (1) the adversary can make a signcryption query in which the sender's private key can generate a valid forgery in unforgeability game; (2) the adversary can make an unsigncryption query in which the receiver's private key can decrypt correctly the challenged ciphertext in confidentiality game; (3) the adversary can make a signature query in which the sender's private key can generate a valid forgery of signcryption ciphertext in unforgeability game; and (4) the adversary can make a decryption query in which the receiver's private key can decrypt correctly the challenged signcrypted ciphertext in confidentiality game. Paterson et al.'s construction [16] is for traditional public key infrastructure (PKI), not identity-based. Subsequently, Chen et al. [17] discussed the attribute-based combined public key problem. They proposed a security model for combined attribute-based encryption (ABE) and attributed-based signature (ABS) based on Waters's scheme. In CT-RSA 2012, Degabriele et al. [18] analyzed the key reuse problem in EMV. They gave an attack method against EMV's RSA algorithms. An adversary can forge a signature about a chosen message by accessing a partial decryption oracle. Vasco et al. [19] discussed the identity-based combined public key problem. They proved that Hess's IBS [5] and Boneh and Franklin's IBE [8] can be simultaneously used. However, they only limit IBS and IBE.

1.1 Contributions

The contribution of our paper is to investigate the identity-based combined public key schemes in two and three cryptographic primitives environments, IBS and IBE environment and IBS, IBE, and IBSC environment. We make the following contributions:

1. We give an identity-based combined signature and encryption (IBCSE) based on Cha and Cheon's IBS [6] and Boneh and Franklin's IBE [8].
2. We define relatively weak but more reasonable security notions for identity-based combined signature, encryption and signcryption (IBCSESC).
3. We give a weakly secure IBCSESC that satisfies our weak security notions and a strongly secure IBCSESC that satisfies strong security notions defined by Paterson et al. [16].

1.2 Organization

Our paper is organized as follows. The security model of IBCSE and an efficient construction are given in Sect. 2. We define relatively weak but more reasonable security notions for IBCSESC in Sect. 3. Two efficient IBCSESC schemes are given in Sect. 4. We describe the conclusions in Sect. 5.

2 Identity-Based Combined Signature and Encryption

Vasco et al. [19] gave the definition for IBCSE. We first review it and give a secure construction.

2.1 Syntax

An IBCSE is a combination of an IBS and an IBE that share a system setup algorithm and a key extraction algorithm.

- **Setup**: is a probabilistic polynomial time (PPT) algorithm performed by a PKG. The PKG first takes as input a security parameter k and then outputs the system public parameters par and a master secret key msk.
- **Extract**: is a key extraction algorithm that first takes as input an identity $ID \in \{0, 1\}^*$, par and msk, and then outputs a private key S_{ID}.
- **Sign**: is a PPT signature algorithm that first takes as input par, a message m, an identity ID, and a private key S_{ID} and then outputs a signature result σ.
- **Verify**: is a deterministic verification algorithm that first takes as input par, a signature σ, a message m, and an identity ID and then outputs a correct symbol $\top$ (σ is a valid signature) or an error symbol $\perp$ (σ is not a valid signature).
- **Encrypt**: is a PPT encryption algorithm that first takes as input par, a plaintext m, and an identity ID, and then outputs a ciphertext c.
- **Decrypt**: is a deterministic decryption algorithm that takes as input par, a ciphertext c, an identity ID, and a private key S_{ID} and outputs a plaintext m or an error $\perp$.

In the above algorithms, (**Setup**, **Extract**, **Sign**, **Verify**) make up an IBS and (**Setup**, **Extract**, **Encrypt**, **Decrypt**) make up an IBE. In addition, these algorithms satisfy the consistency condition, i.e., if $\sigma = \mathbf{Sign}(par, m, ID, S_{ID})$, then $\top = \mathbf{Verify}(par, \sigma, m, ID)$ and if $c = \mathbf{Encrypt}(par, m, ID)$, then $m = \mathbf{Decrypt}(par, c, ID, S_{ID})$.

2.2 Security Notions

The standard security notion for IBS is existential unforgeability against adaptively chosen messages attacks under chosen identity attacks (EUF-ID-CMA). The standard acceptable security notion for IBE is indistinguishability against adaptive chosen ciphertext attacks under chosen identity attacks (IND-ID-CCA2). For combined public key schemes, we should permit the same identity to be applied for both encryption and signature. That is, the same key pair is permitted to decrypt a ciphertext and sign a message. To reflect such ability of adversary, we need to extend standard EUF-ID-CMA and IND-ID-CCA2 security notions. For EUF-ID-CMA security of an IBS, we should allow the adversary to access not only signature oracle but also decryption oracle since the same private key will be applied to decrypt a ciphertext. Similarly, for IND-ID-CCA2 security of an IBE, we should allow the adversary to access not only decryption oracle but also signature oracle since the same private key will be applied to sign a message. The formal security notions for IBCSE are showed below.

The EUF-ID-CMA of the signature component of an IBCSE is defined through the following game.

- **Initial**: A challenger $\mathcal{C}$ first performs the **Setup** algorithm with a security parameter k and then gives the system parameters par to an adversary $\mathcal{F}$.
- **Attack**: $\mathcal{F}$ adaptively performs a polynomially bounded number of queries.
  - Key extraction queries: Given an identity ID by $\mathcal{F}$, $\mathcal{C}$ returns a private key $S_{ID} = \mathbf{Extract}(par, msk, ID)$ to $\mathcal{F}$.
  - Signature queries: Given an identity ID and a message m by $\mathcal{F}$, $\mathcal{C}$ first computes $S_{ID} = \mathbf{Extract}(par, msk, ID)$ and then returns the result of $\mathbf{Sign}(par, m, ID, S_{ID})$ to $\mathcal{F}$.
  - Decryption queries: Given an identity ID and a ciphertext c by $\mathcal{F}$, $\mathcal{C}$ first computes $S_{ID} = \mathbf{Extract}(par, msk, ID)$ and then returns the result of $\mathbf{Decrypt}(par, c, ID, S_{ID})$ to $\mathcal{F}$ (this result may be an error symbol $\perp$ when c is not valid).
- **Forgery**: $\mathcal{F}$ outputs a message m^*, an identity ID^*, and a signature σ^*. If the following conditions hold, $\mathcal{F}$ succeeds in this game.

1. $\mathbf{Verify}(par, \sigma^*, m^*, ID^*) = \top$.
2. ID^* has not been submitted to a key extraction query.
3. (m^*, ID^*) has not been submitted to a signature query.

The probability that $\mathcal{F}$ wins is defined as its advantage, i.e.,

$$Adv^{\text{EUF-ID-CMA}}_{\text{IBCSE-IBS}}(\mathcal{F}) = \Pr[\mathcal{F}\ wins].$$

In the above game, $\mathcal{F}$ can ask a decryption query on (c, ID^*). Of course, $\mathcal{F}$ can make both signature oracle and decryption oracle on the same identity ID. This is the main difference between IBCSE and standard IBS.

Definition 1 If an adversary has advantage at least ϵ in the above game, runs in time at most t, and asks at most q_k key extraction queries, q_s signature queries and q_d decryption queries, we say that it is an $(\epsilon, t, q_k, q_s, q_d)$-forger of the signature component of an IBCSE. If no $(\epsilon, t, q_k, q_s, q_d)$-forger exists, we say that the signature component of an IBCSE is $(\epsilon, t, q_k, q_s, q_d)$-EUF-ID-CMA.

The IND-ID-CCA2 of the encryption component of an IBCSE is defined through the following game.

- **Initial**: A challenger $\mathcal{C}$ performs the **Setup** algorithm with a security parameter k and gives the system public parameters *par* to an adversary $\mathcal{A}$.
- **Phase 1**: $\mathcal{A}$ asks a polynomially bounded number of queries just like in the EUF-ID-CMA game of the signature component of an IBCSE.
- **Challenge**: $\mathcal{A}$ determines when the Phase 1 finishes. $\mathcal{A}$ produces two challenged messages (m_0, m_1) with the same length and an identity ID^*. The ID^* was not submitted to a key extraction query in Phase 1. $\mathcal{C}$ chooses a random bit γ from $\{0, 1\}$ and sends $c^* = \mathbf{Encrypt}(par, m_\gamma, ID^*)$ to $\mathcal{A}$.
- **Phase 2**: $\mathcal{A}$ adaptively asks a polynomially bounded number of queries again as in Phase 1 with the following restrictions:

1. $\mathcal{A}$ cannot ask a key extraction query on ID^*.
2. $\mathcal{A}$ cannot ask a decryption query on (c^*, ID^*).

- **Guess**: $\mathcal{A}$ outputs a bit γ' and succeeds in this game if $\gamma' = \gamma$.

The advantage of $\mathcal{A}$ is defined to be

$$Adv^{\text{IND-ID-CCA2}}_{\text{IBCSE-IBE}}(\mathcal{A}) = |2\Pr[\gamma' = \gamma] - 1|.$$

In the above game, $\mathcal{A}$ can ask a signature query on (m, ID^*). Of course, $\mathcal{A}$ can ask both signature oracle and decryption oracle on the same identity *ID*. This is the main difference between IBCSE and standard IBE.

Definition 2 If an adversary $\mathcal{A}$ has advantage at least ϵ in the above game, runs in time at most t, and makes at most q_k key extraction queries, q_s signature queries and q_d decryption queries, we say that it is an $(\epsilon, t, q_k, q_s, q_d)$-attacker of the encryption component of an IBCSE. If no $(\epsilon, t, q_k, q_s, q_d)$-attacker exists, we say that the encryption component of an IBCSE is $(\epsilon, t, q_k, q_s, q_d)$-IND-ID-CCA2.

Definition 3 An IBCSE is said to be jointly secure if its signature component is EUF-ID-CMA and its encryption component is IND-ID-CCA2.

2.3 A Secure IBCSE Scheme

We use Cha and Cheon's IBS [6] and Boneh and Franklin's IBE [8] to construct a secure IBCSE.

- **Setup**: Given a security parameter k, the PKG selects a cyclic additive groups G_1 and a cyclic multiplicative G_2 with the same prime order q, a generator P of G_1, a bilinear map $\hat{e} : G_1 \times G_1 \to G_2$, and five hash functions $H_1 : \{0, 1\}^* \to G_1, H_2 : \{0, 1\}^n \times G_1 \to \mathbb{Z}_q^*, H_3 : \{0, 1\}^n \times \{0, 1\}^n \to \mathbb{Z}_q^*, H_4 : G_2 \to$

$\{0, 1\}^n$, and $H_5 : \{0, 1\}^n \rightarrow \{0, 1\}^n$. Here, n is the length of a message to be signed or encrypted. It selects a master key $s \in \mathbb{Z}_q^*$ and publishes system public parameters $par = \{G_1, G_2, n, \hat{e}, P, P_{pub}, H_1, H_2, H_3, H_4, H_5\}$, where $P_{pub} = sP$. The master key s should be kept secret.

- **Extract**: Given a user's identity $ID \in \{0, 1\}^*$, the PKG sets the user's private key $S_{ID} = sQ_{ID}$, where $Q_{ID} = H_1(ID)$.
- **Sign**: To sign a message $m \in \{0, 1\}^n$, the signer does the steps below.

1. Select r from $\mathbb{Z}_q^*$ randomly.
2. Compute $X = rQ_{ID}$.
3. Compute $h = H_2(m, X)$.
4. Compute $Z = (r + h)S_{ID}$.
5. The signature is $\sigma = (X, Z)$.

- **Verify**: Given a signature $\sigma = (X, Z)$ of a message m for an identity ID, any third party does the steps below.

1. Compute $h = H_2(m, X)$.
2. Accept this signature if and only if $\hat{e}(Z, P) = \hat{e}(P_{pub}, X + hQ_{ID})$. Otherwise, return $\perp$.

- **Encrypt**: To encrypt $m \in \{0, 1\}^n$, a sender does the steps below.

1. Select $\lambda \in \{0, 1\}^n$ randomly.
2. Set $t = H_3(\lambda, m)$.
3. Compute $U = tP$, $V = \lambda \oplus H_4(\hat{e}(Q_{ID}, P_{pub})^t)$ and $W = m \oplus H_5(\lambda)$.
4. The ciphertext is $c = (U, V, W)$.

- **Decrypt**: Given a ciphertext $c = (U, V, W)$, the receiver does the steps below.

1. Compute $\lambda = V \oplus H_4(\hat{e}(U, S_{ID}))$.
2. Compute $m = W \oplus H_5(\lambda)$.
3. Set $t = H_3(\lambda, m)$.
4. Check if $U = tP$ holds. If yes, output the message m. Otherwise, output an error $\perp$.

In Cha and Cheon's IBS [6], the system public parameters are

$$par_{cc} = \{G_1, G_2, n, \hat{e}, P, P_{pub}, H_1, H_2\}.$$

In Boneh and Franklin's IBE [8], the system public parameters are

$$par_{bf} = \{G_1, G_2, n, \hat{e}, P, P_{pub}, H_1, H_3, H_4, H_5\}.$$

That is, the two schemes share the same parameters $\{G_1, G_2, n, \hat{e}, P, P_{pub}, H_1\}$. The two schemes also share the same **Extract** algorithm. Now, we give the security result for such construction method in Theorem 1.

Theorem 1 *The above IBCSE is jointly secure in the random oracle model.*

Proof We point out that the signature component of above IBCSE is EUF-ID-CMA and its encryption component is IND-ID-CCA2. The proof proceeds by reductionist proof.

On the one hand, we prove that if an adversary $\mathcal{F}$ can break the $(\epsilon, t, q_k, q_s, q_d)$-EUF-ID-CMA of above IBCSE, then one can construct an algorithm $\mathcal{C}$ that can break the (ϵ, t, q_k, q_s)-EUF-ID-CMA of Cha and Cheon's IBS.

- **Initial**: Given Cha and Cheon's IBS system parameters par_{cc}, $\mathcal{C}$ chooses H_3, H_4, and H_5 and sends $par = \{G_1, G_2, n, \hat{e}, P, P_{pub}, H_1, H_2, H_3, H_4, H_5\}$ to $\mathcal{F}$.
- **Attack**: When $\mathcal{F}$ asks a key extraction query on *ID*, $\mathcal{C}$ submits *ID* to its key extraction oracle and gives the result to $\mathcal{F}$. Similarly, when $\mathcal{F}$ makes a signature query on (m, ID), $\mathcal{C}$ submits (m, ID) to its signature oracle and gives the result to $\mathcal{F}$. The important problem is how to answer the decryption queries. To do this, H_3 is regarded as a random oracle. $\mathcal{C}$ keeps a list L_3 to record all queries and answers of H_3 (the consistency will be maintained and collision will be avoided). When $\mathcal{F}$ asks a decryption query on (c, ID), $\mathcal{C}$ checks if L_3 already includes an entry (λ, m, t) such that $U = tP$, $V = \lambda \oplus H_4(\hat{e}(Q_{ID}, P_{pub})^t)$ and $W = m \oplus H_5(\lambda)$. If yes, the answer to the decryption query is m. Otherwise, $\mathcal{C}$ returns an error $\perp$. Of course, the other hash functions are also regarded as random oracles. Here, we only emphasize H_3 because it is important to answer the decryption queries.
- **Forgery**: $\mathcal{F}$ outputs a message m^*, an identity ID^* and a signature σ^*. If $\mathcal{F}$'s forgery is valid, then $\mathcal{C}$ also can output a valid forgery (m^*, ID^*, σ^*). Here, we require that $\mathcal{F}$ has not asked a key extraction query on ID^*. $\mathcal{C}$ never queried ID^* to its key extraction oracle. $\mathcal{F}$ has not asked a signature query on (m^*, ID^*). $\mathcal{C}$ also never queried (m^*, ID^*) to its signature oracle. Therefore, $\mathcal{C}$ succeeds with the same probability as $\mathcal{F}$.

On the other hand, if an adversary $\mathcal{A}$ can break the $(\epsilon, t, q_k, q_s, q_d)$-IND-ID-CCA2 of above IBCSE, then one can construct an algorithm $\mathcal{C}$ that can break (ϵ, t, q_k, q_d)-IND-ID-CCA2 of Boneh and Franklin's IBE.

- **Initial**: Given Boneh and Franklin's IBE system public parameters par_{bf}, $\mathcal{C}$ chooses H_2 and sends $par = \{G_1, G_2, n, \hat{e}, P, P_{pub}, H_1, H_2, H_3, H_4, H_5\}$ to $\mathcal{A}$.
- **Phase 1**: When $\mathcal{A}$ asks a key extraction query on *ID*, $\mathcal{C}$ submits *ID* to its key extraction oracle and gives the result to $\mathcal{A}$. Similarly, when $\mathcal{A}$ asks a decryption query on (c, ID), $\mathcal{C}$ submits (c, ID) to its decryption oracle and gives the answer to $\mathcal{A}$. The important problem is how to answer the signature queries. To do this, H_2 is regarded as a random oracle. When $\mathcal{A}$ asks a signature query on (m, ID), $\mathcal{C}$ randomly selects $r, h \in \mathbb{Z}_q^*$, computes $X = rP - hQ_{ID}$ and $Z = rP_{pub}$, and sets the hash value $H_2(m, X)$ to be h. The answer to the signature query is $\sigma = (X, Z)$.
- **Challenge**: $\mathcal{A}$ generates two challenged messages (m_0, m_1) with the same length and an identity ID^*. ID^* has not been submitted to a key extraction query in Phase 1. $\mathcal{C}$ outputs the same messages (m_0, m_1) and identity ID^* as its challenge. $\mathcal{C}$ receives the challenged ciphertext c^* and transfers to $\mathcal{A}$.

- **Phase 2**: $\mathcal{C}$ deals with queries as in Phase 1. $\mathcal{A}$ cannot ask a key extraction query on ID^* and a decryption query on (c^*, ID^*). $\mathcal{C}$ also does not submit ID^* to its key extraction oracle and (c^*, ID^*) to its decryption oracle.
- **Guess**: $\mathcal{A}$ finally outputs a guess γ' which $\mathcal{C}$ outputs the same value as its guess. $\mathcal{C}$ succeeds with the same probability as $\mathcal{A}$. □

From Theorem 1, we know that Cha and Cheon's IBS and Boneh and Franklin's IBE can be securely used with the same key pair.

3 Identity-Based Combined Signature, Encryption and Signcryption

We give a formal model for identity-based combined signature, encryption and signcryption.

3.1 Syntax

An IBCSESC is a combination of an IBS, an IBE, and an IBSC that share a system setup algorithm and a key extraction algorithm. It consists of **Setup**, **Extract**, **Sign**, **Verify**, **Encrypt**, **Decrypt**, **Signcrypt**, and **Unsigncrypt**. The first six algorithms are the same as in Sect. 2.1, and the other algorithms are described as follows.

- **Signcrypt**: is a PPT signcryption algorithm which takes as input par, a message m, a sender's identity ID_s, a sender's private key S_{ID_s}, and a receiver's identity ID_r, and outputs a ciphertext c.
- **Unsigncrypt**: is a deterministic unsigncryption algorithm that takes as input par, a ciphertext c, a sender's identity ID_s, a receiver's identity ID_r, and a receiver's private key S_{ID_r}, and outputs a plaintext m or an error $\perp$ (c is not valid).

In an IBCSESC, (**Setup**, **Extract**, **Sign**, **Verify**) make up an IBS, (**Setup**, **Extract**, **Encrypt**, **Decrypt**) make up an IBE, and (**Setup**, **Extract**, **Signcrypt**, **Unsigncrypt**) make up an IBSC. These algorithms must satisfy the consistency constraint for signature, encryption, and signcryption. The consistency constraint for signature and encryption is the same as in Sect. 2.1. The consistency constrain for signcryption is that if $c = \textbf{Signcrypt}(par, m, ID_s, S_{ID_s}, ID_r)$, then $m = \textbf{Unsigncrypt}(par, c, ID_s, ID_r, S_{ID_r})$.

3.2 Security Notions

Since an IBCSESC has three components, an IBS, an IBE, and an IBSC, we should define the corresponding security notions for the three components. We first define relatively weak but more reasonable security notions for an IBCSESC.

The weak EUF-ID-CMA of the signature component of an IBCSESC is defined through the following game.

- **Initial**: A challenger $\mathcal{C}$ runs the **Setup** algorithm with a security parameter k and gives the system public parameters par to an adversary $\mathcal{F}$.
- **Attack**: $\mathcal{F}$ adaptively performs a polynomially bounded number of queries.
  - Key extraction queries: Given an identity ID by $\mathcal{F}$, $\mathcal{C}$ sends the private key $S_{ID} = \mathbf{Extract}(par, msk, ID)$ to $\mathcal{F}$.
  - Signature queries: Given an identity ID and a message m by $\mathcal{F}$, $\mathcal{C}$ computes $S_{ID} = \mathbf{Extract}(par, msk, ID)$ and sends the output result of **Sign** (par, m, ID, S_{ID}) to $\mathcal{F}$.
  - Decryption queries: Given an identity ID and a ciphertext c by $\mathcal{F}$, $\mathcal{C}$ computes $S_{ID} = \mathbf{Extract}(par, msk, ID)$ and sends the output result of $\mathbf{Decrypt}(par, c, ID, S_{ID})$ to $\mathcal{F}$ (this result may be an error $\perp$ if c is not valid).
  - Signcryption queries: Given a sender's identity ID_s, a receiver's identity ID_r, and a message m by $\mathcal{F}$, $\mathcal{C}$ computes $S_{ID_s} = \mathbf{Extract}(par, msk, ID_s)$ and sends $c = \mathbf{Signcrypt}(par, m, ID_s, S_{ID_s}, ID_r)$ to $\mathcal{F}$.
  - Unsigncryption queries: Given a sender's identity ID_s, a receiver's identity ID_r, and a ciphertext c by $\mathcal{F}$, $\mathcal{C}$ computes $S_{ID_r} = \mathbf{Extract}(par, msk, ID_r)$ and sends the output of $\mathbf{Unsigncrypt}(par, c, ID_s, ID_r, S_{ID_r})$ to $\mathcal{F}$ (this output may be an error $\perp$ if c is not valid).
- **Forgery**: $\mathcal{F}$ finally outputs a message m^*, an identity ID^*, and a signature σ^*. $\mathcal{F}$ succeeds if the following conditions hold:

1. $\mathbf{Verify}(par, \sigma^*, m^*, ID^*) = \top$.
2. ID^* has not been submitted to a key extraction query.
3. (m^*, ID^*) has not been submitted to a signature query.
4. $\mathcal{F}$ has not asked a signcryption query on (m^*, ID^*, ID_r). Here, ID_r is an arbitrary receiver's identity.

The probability that $\mathcal{F}$ wins is defined as its advantage, i.e.,

$$Adv^{\text{EUF-ID-CMA}}_{\text{IBCSESC-IBS}}(\mathcal{F}) = \Pr[\mathcal{F}\ wins].$$

Note that there is a restriction in forgery phase that $\mathcal{F}$ has not asked a signcryption query on (m^*, ID^*, ID_r). However, there is no such a restriction in Paterson et al.'s definition [16]. If there is no such a restriction, we call strong EUF-ID-CMA of the signature component of an IBCSESC.

Definition 4 If $\mathcal{F}$ has advantage at least ϵ in the above weak game (resp. strong game), runs in time at most t, and makes at most q_k key extraction queries, q_s signature queries, q_d decryption queries, q_{sc} signcryption queries, and q_{usc} unsigncryption queries, we say that $\mathcal{F}$ is a weak (resp. strong) $(\epsilon, t, q_k, q_s, q_d, q_{sc}, q_{usc})$-forger of the signature component of an IBCSESC. If no weak (resp. strong) $(\epsilon, t, q_k, q_s, q_d, q_{sc}, q_{usc})$-forger exists, we say that the signature component of an IBCSESC is weak (resp. strong) $(\epsilon, t, q_k, q_s, q_d, q_{sc}, q_{usc})$-EUF-ID-CMA.

The weak IND-ID-CCA2 of the encryption component of an IBCSESC is defined through the following game.

- **Initial**: A challenger $\mathcal{C}$ runs the **Setup** algorithm with a security parameter k and gives the system public parameters *par* to an adversary $\mathcal{A}$.
- **Phase 1**: $\mathcal{A}$ asks a polynomially bounded number of queries just like in the EUF-ID-CMA game of the signature component of an IBCSESC.
- **Challenge**: $\mathcal{A}$ determines when Phase 1 ends. $\mathcal{A}$ generates two challenged messages (m_0, m_1) with the same length and an identity ID^*. ID^* has not been submitted to a key extraction query in Phase 1. $\mathcal{C}$ selects a random bit γ from $\{0, 1\}$ and sends $c^* = \textbf{Encrypt}(par, m_\gamma, ID^*)$ to $\mathcal{A}$.
- **Phase 2**: $\mathcal{A}$ can adaptively ask a polynomially bounded number of queries again as in Phase 1 with the following restrictions:

1. ID^* will not be submitted to a key extraction query.
2. (c^*, ID^*) will not be submitted to a decryption query.
3. It cannot ask an unsigncryption query on $(c^*(\text{or variations of } c^*), ID_s, ID^*)$ to obtain the plaintext. Here, ID_s is an arbitrary sender's identity.

- **Guess**: $\mathcal{A}$ finally outputs a bit γ' and succeeds in this game if $\gamma' = \gamma$.

The advantage of $\mathcal{A}$ is defined to be

$$Adv_{\text{IBCSESC-IBE}}^{\text{IND-ID-CCA2}}(\mathcal{A}) = |2\Pr[\gamma' = \gamma] - 1|.$$

Note that there is a restriction in Phase 2 that $\mathcal{A}$ cannot ask an unsigncryption query on $(c^*(\text{or variations of } c^*), ID_s, ID^*)$ to obtain the plaintext. There is no such a restriction in Paterson et al.'s definition [16]. If there is no such a restriction, we call strong IND-ID-CCA2 of the encryption component of an IBCSESC.

Definition 5 If $\mathcal{A}$ has advantage at least ϵ in the above weak game (resp. strong game), runs in time at most t, and makes at most q_k key extraction queries, q_s signature queries, q_d decryption queries, q_{sc} signcryption queries, and q_{usc} unsigncryption queries, we say that $\mathcal{A}$ is a weak (resp. strong) $(\epsilon, t, q_k, q_s, q_d, q_{sc}, q_{usc})$-attacker of the encryption component of an IBCSESC. If no weak (resp. strong) $(\epsilon, t, q_k, q_s, q_d, q_{sc}, q_{usc})$-attacker exists, we say that the encryption component of an IBCSESC is weak (resp. strong) $(\epsilon, t, q_k, q_s, q_d, q_{sc}, q_{usc})$-IND-ID-CCA2.

The weak EUF-ID-CMA of the signcryption component of an IBCSESC is defined through the following game.

- **Initial**: A challenger $\mathcal{C}$ runs the **Setup** algorithm with a security parameter k and gives the system public parameters *par* to an adversary $\mathcal{F}$.
- **Attack**: $\mathcal{F}$ can perform a polynomially bounded number of queries just like in the EUF-ID-CMA game of the signature component of an IBCSESC.

- **Forgery**: $\mathcal{F}$ finally produces a ciphertext c^*, a sender's identity ID_s^*, and a receiver's identity ID_r^*. When the following conditions hold, $\mathcal{F}$ wins this game.

1. **Unsigncrypt**$(par, c^*, ID_s^*, ID_r^*, S_{ID_r^*}) = m^*$.
2. ID_s^* has not been submitted to a key extraction query.
3. (m^*, ID_s^*) has not been submitted to a signature query.
4. $\mathcal{F}$ has not made a signcryption query on (m^*, ID_s^*, ID_r^*).

The probability that $\mathcal{F}$ wins is defined as its advantage, i.e.,

$$Adv_{\text{IBCSESC-IBSC}}^{\text{EUF-ID-CMA}}(\mathcal{F}) = \Pr[\mathcal{F}\ wins].$$

Note that there is a restriction in forgery phase that $\mathcal{F}$ cannot make a signature query on (m^*, ID_s^*). There is no such a restriction in Paterson et al.'s definition [16]. If there is no such a restriction, we call strong EUF-ID-CMA security of the signcryption component of an IBCSESC.

Definition 6 If $\mathcal{F}$ has advantage at least ϵ in the above weak game (resp. strong game), runs in time at most t, and makes at most q_k key extraction queries, q_s signature queries, q_d decryption queries, q_{sc} signcryption queries, and q_{usc} unsigncryption queries, we say that $\mathcal{F}$ is a weak (resp. strong) $(\epsilon, t, q_k, q_s, q_d, q_{sc}, q_{usc})$-forger of the signcryption component of an IBCSESC. If no weak (resp. strong) $(\epsilon, t, q_k, q_s, q_d, q_{sc}, q_{usc})$-forger exists, we say that the signcryption component of an IBCSESC is weak (resp. strong) $(\epsilon, t, q_k, q_s, q_d, q_{sc}, q_{usc})$-EUF-ID-CMA.

The weak IND-ID-CCA2 of the signcryption component of an IBCSESC is defined through the following game.

- **Initial**: A challenger $\mathcal{C}$ runs the **Setup** algorithm with a security parameter k and gives the system public parameters *par* to an adversary $\mathcal{A}$.
- **Phase 1**: $\mathcal{A}$ asks a polynomially bounded number of queries just like in the EUF-ID-CMA game of the signature component of an IBCSESC.
- **Challenge**: $\mathcal{A}$ determines when Phase 1 ends. $\mathcal{A}$ generates two challenged plaintexts (m_0, m_1) with the same length, a sender's identity ID_s^*, and a receiver's identity ID_r^*. Here, we require that ID_r^* has not been submitted to a key extraction query in Phase 1. $\mathcal{C}$ selects a random bit γ from $\{0, 1\}$ and sends $c^* =$ **Signcrypt**$(par, m_\gamma, ID_s^*, S_{ID_s^*}, ID_r^*)$ to $\mathcal{A}$.
- **Phase 2**: $\mathcal{A}$ can adaptively ask a polynomially bounded number of queries again as in Phase 1 with the following restrictions:

1. ID_r^* is not submitted to a key extraction query.
2. It cannot ask a decryption query on $(c^*(\text{or variations of } c^*), ID_r^*)$ to obtain the plaintext.
3. It cannot ask an unsigncryption query on (c^*, ID_s^*, ID_r^*) to obtain the plaintext.

- **Guess**: $\mathcal{A}$ finally outputs a bit γ' and wins this game if $\gamma' = \gamma$.

The advantage of $\mathcal{A}$ is defined to be

$$Adv_{\text{IBCSESC-IBSC}}^{\text{IND-ID-CCA2}}(\mathcal{A}) = |2\Pr[\gamma' = \gamma] - 1|.$$

Note that there is a restriction in Phase 2 that $\mathcal{A}$ cannot ask a decryption query on $(c^*(\text{or variations of } c^*), ID_r^*)$ to obtain the plaintext. There is no such a restriction in Paterson et al.'s definition [16]. If there is no such a restriction, we call strong IND-ID-CCA2 of the signcryption component of an IBCSESC.

Definition 7 If $\mathcal{A}$ has advantage at least ϵ in the above weak game (resp. strong game), runs in time at most t, and makes at most q_k key extraction queries, q_s signature queries, q_d decryption queries, q_{sc} signcryption queries, and q_{usc} unsigncryption queries, we say that $\mathcal{A}$ is a weak (resp. strong) $(\epsilon, t, q_k, q_s, q_d, q_{sc}, q_{usc})$-attacker of the signcryption component of an IBCSESC. If no weak (resp. strong) $(\epsilon, t, q_k, q_s, q_d, q_{sc}, q_{usc})$-attacker exists, we say that the signcryption component of an IBCSESC is weak (resp. strong) $(\epsilon, t, q_k, q_s, q_d, q_{sc}, q_{usc})$-IND-ID-CCA2.

Definition 8 An IBCSESC is said to be weak (resp. strong) jointly secure if its signature component is weak (resp. strong) EUF-ID-CMA, its encryption component is weak (resp. strong) IND-ID-CCA2, and its signcryption component is weak (resp. strong) EUF-ID-CMA and weak (resp. strong) IND-ID-CCA2.

Now, we explain why Paterson et al.'s definition [16] is too strong. A signcryption scheme is usually constructed from some signature and encryption schemes. A signcrypted ciphertext must contain some elements of signature and encryption ciphertext. To break the security of signature and encryption components, the adversary can make signcryption and unsigncryption queries, respectively. To break the security of signcryption component, the adversary can make signature and decryption queries. Therefore, it is difficult for a combined signature, encryption and signcryption scheme to satisfy strong security. Therefore, it is more reasonable that we require that a combined signature, encryption and signcryption scheme only satisfies weak security. To achieve the strong security, we cannot use a simple combination of a signature, an encryption, and a signcryption that share a system setup algorithm and a key generation algorithm. We must use different structures in signature, encryption, and signcryption components so that (1) a signcryption query is not useful for a forger of signature component; (2) an unsigncryption query is not useful for an attacker of encryption component; (3) a signature query is not useful for a forger of signcryption component; and (4) a decryption query is not useful for an attacker of signcryption component.

4 Two IBCSESC Constructions

In this section, we give two IBCSESC constructions. The first one satisfies the weak security, and the second one satisfies the strong security.

4.1 A Weakly Secure IBCSESC Scheme

An IBCSESC can be constructed by independent IBS, IBE, and IBSC. Here, we give a example using Cha and Cheon's IBS [6], Boneh and Franklin's IBE [8], and Chen and Malone-Lee's IBSC [11]. The detail is described as follows. **Setup**, **Extract**, **Sign**, **Verify**, **Encrypt**, and **Decrypt** algorithms are the same as in Sect. 2.3 except that we need another hash function $H_6 : G_2 \rightarrow \{0, 1\}^{|G_1|+|ID|+n}$. Here, $|G_1|$ and $|ID|$ are the numbers of bits for an element of G_1 and an identity ID, respectively. **Signcrypt** and **Unsigncrypt** algorithms are described below.

- **Signcrypt**: To send a message $m \in \{0, 1\}^n$ to a receiver with identity ID_r, a sender with identity ID_s does the steps below.

1. Choose $r \in \mathbb{Z}_q^*$ randomly.
2. Compute $X = rQ_{ID_s}$, $h = H_2(m, X)$ and $Z = (r + h)S_{ID_s}$.
3. Compute $\omega = \hat{e}(rS_{ID_s}, Q_{ID_r})$ and $y = H_6(\omega) \oplus (Z||ID_s||m)$.
4. The ciphertext is $c = (X, y)$.

- **Unsigncrypt**: Given a ciphertext $c = (X, y)$, the receiver does the steps below.

1. Compute $\omega = \hat{e}(X, S_{ID_r})$ and $Z||ID_s||m = y \oplus H_6(\omega)$.
2. Compute $h = H_2(m, X)$.
3. Verify if $\hat{e}(Z, P) = \hat{e}(P_{pub}, X + hQ_{ID})$ holds. If this equation holds, output the message m. Otherwise, output an error $\perp$.

This simple construction is only weakly secure, not strongly secure. As an example, we explain how to break the strong EUF-ID-CMA in the signature component of the above IBCSESC. To forge a valid signature of message m^* on identity ID^*, $\mathcal{F}$ follows the steps below.

1. Choose an identity ID_r and make a signcryption query on (m^*, ID^*, ID_r) to get a ciphertext $c = (X, y)$.
2. Ask a key extraction query on ID_r to get the private key S_{ID_r}.
3. Compute $\omega = \hat{e}(X, S_{ID_r})$ and $Z||ID_s||m = y \oplus H_6(\omega)$.
4. Output a forgery $\sigma^* = (X, Z)$.

It is obvious that $\sigma^* = (X, Z)$ is valid for the message m^* and the identity ID^*. Therefore, the above IBCSESC is not strongly secure. The proof that the above IBCSESC satisfies the weak security is similar to the proof in Theorem 1. We do not give the detail.

4.2 A Strongly Secure IBCSESC Scheme

We propose a strongly secure IBCSESC based on an IBCSE. To achieve the strong security, we use different tags in signature, encryption, and signcryption components. Given an IBCSE $\mathcal{IBCSE}$ = ($\mathcal{IBCSE}$.**Setup**, $\mathcal{IBCSE}$.**Extract**, $\mathcal{IBCSE}$.**Sign**,

$\mathcal{IBCSESC}$.**Setup:** On input a security parameter k:
1. $(par, msk) = \mathcal{IBCSE}$.**Setup**$(k)$
2. Output the system public parameters par and the master secret key msk

$\mathcal{IBCSESC}$.**Extract:** On input par, msk and an identity ID:
1. $S_{ID} = \mathcal{IBCSE}$.**Extract**(par, msk, ID)
2. Output the private key S_{ID}

$\mathcal{IBCSESC}$.**Sign:** On input par, a message m, an identity ID and a private key S_{ID}:
1. $\sigma = \mathcal{IBCSE}$.**Sign**$(par, 0||m, ID, S_{ID})$
2. Output the signature σ

$\mathcal{IBCSESC}$.**Verify:** On input par, a signature σ, a message m and an identity ID:
1. $\top$ or $\bot = \mathcal{IBCSE}$.**Verify**$(par, \sigma, 0||m, ID)$
2. Output $\top$ for acceptance or $\bot$ for rejection

$\mathcal{IBCSESC}$.**Encrypt:** On input par, a message m and an identity ID:
1. $tag = 0$
2. $c = \mathcal{IBCSE}$.**Encrypt**$(par, tag||m, ID)$
3. Output the ciphertext c

$\mathcal{IBCSESC}$.**Decrypt:** On input par, a ciphertext c, an identity ID and a private key S_{ID}:
1. $m' = \mathcal{IBCSE}$.**Decrypt**(par, c, ID, S_{ID})
2. If $m' = \bot$, return $\bot$. Otherwise, perform the following steps
3. $tag||m = m'$
4. If $tag = 0$, output the message m. Otherwise output $\bot$

$\mathcal{IBCSESC}$.**Signcrypt:** On input par, a message m, a sender's identity ID_s, a sender's privatekey S_{ID_s} and a receiver's identity ID_r:
1. $\sigma = \mathcal{IBCSE}$.**Sign**$(par, 1||ID_r||m, ID_s, S_{ID_s})$
2. $tag = 1||ID_s$
3. $c = \mathcal{IBCSE}$.**Encrypt**$(par, tag||m||\sigma, ID_r)$
4. Output the ciphertext c

$\mathcal{IBCSESC}$.**Unsigncrypt:** On input par, a ciphertext c, a sender's identity ID_s, a receiver's identity ID_r and a receiver's private key S_{ID_r}:
1. $m' = \mathcal{IBCSE}$.**Decrypt**(par, c, ID_r, S_{ID_r})
2. If $m' = \bot$, return $\bot$. Otherwise, perform the following steps
3. $tag||m||\sigma = m'$
4. $tag = 1||ID_s$
5. If $\top = \mathcal{IBCSE}$.**Verify**$(par, \sigma, 1||ID_r||m, ID_s)$, output the message m. Otherwise output $\bot$

Fig. 1 A strongly secure IBCSESC scheme

$\mathcal{IBCSE}$.**Verify**, $\mathcal{IBCSE}$.**Decrypt**), a strongly secure IBCSESC $\mathcal{IBCSESC}$ is constructed in Fig. 1.

The security for such construction is given in Theorems 2, 3, 4, and 5.

Theorem 2 *The signature part of $\mathcal{IBCSESC}$ is strongly $(\epsilon, t', q_k, q'_s, q'_d, q'_{sc}, q'_{usc})$-EUF-ID-CMA in the presence of additional oracles if the signature component of $\mathcal{IBCSE}$ is $(\epsilon, t, q_k, q_s, q_d)$-EUF-ID-CMA in the presence of a decryption oracle, where $q_s = q'_s + q'_{sc}$, $q_d = q'_d + q'_{usc}$, $t = t' + O(q'_{sc}T_e + q'_{usc}T_v)$. Here, T_e and T_v are the maximum time for performing an encryption and verifying a signature.*

Proof See the Appendix 1. □

Theorem 3 *The encryption part of $\mathcal{IBCSESC}$ is $(\epsilon, t', q_k, q'_s, q'_d, q'_{sc}, q'_{usc})$-IND-ID-CCA2 in the presence of additional oracles if the encryption component of $\mathcal{IBCSE}$ is $(\epsilon, t, q_k, q_s, q_d)$-IND-ID-CCA2 in the presence of a signature oracle, where $q_s = q'_s + q'_{sc}$, $q_d = q'_d + q'_{usc}$, $t = t' + O(q'_{sc}T_e + q'_{usc}T_v)$. Here, T_e and T_v are the maximum time for performing an encryption and verifying a signature.*

Proof See the Appendix 2. □

Theorem 4 *The signcryption part of $\mathcal{IBCSESC}$ is strong $(\epsilon, t', q_k, q'_s, q'_d, q'_{sc}, q'_{usc})$-EUF-ID-CMA in the presence of additional oracles if the signature component of $\mathcal{IBCSE}$ is $(\epsilon, t, q_k, q_s, q_d)$-EUF-ID-CMA in the presence of a decryption oracle, where $q_s = q'_s + q'_{sc}$, $q_d = q'_d + q'_{usc}$, $t = t' + O(q'_{sc}T_e + q'_{usc}T_v)$. Here, T_e and T_v are the maximum time for performing an encryption and verifying a signature.*

Proof See the Appendix 3. □

Theorem 5 *The signcryption part of $\mathcal{IBCSESC}$ is strong $(\epsilon, t', q_k, q'_s, q'_d, q'_{sc}, q'_{usc})$-IND-ID-CCA2 in the presence of additional oracles if the encryption component is $(\epsilon, t, q_k, q_s, q_d)$-IND-ID-CCA2 in the presence of a signature oracle, where $q_s = q'_s + q'_{sc}$, $q_d = q'_d + q'_{usc}$, $t = t' + O(q'_{sc}T_e + q'_{usc}T_v)$. Here, T_e and T_v are the maximum time for performing an encryption and verifying a signature.*

Proof See the Appendix 4. □

5 Conclusions

In this paper, we showed that Cha and Cheon's IBS and Boneh and Franklin's IBE can be applied with the same key pair. We gave relatively weak but more reasonable security notions for IBCSESC. A weakly secure IBCSESC that satisfies our weak security notions and a strongly secure IBCSESC that satisfies strong security notions were proposed, respectively. A future work is to investigate if it is secure to apply the same key pair in signature, encryption, signcryption, and key agreement since a key agreement protocol is a kind of basic cryptographic primitive in information security.

Acknowledgements We would like to thank the anonymous reviewers for their valuable comments and suggestions. This work is supported by the National Natural Science Foundation of China (Grant No 61272525) and the Fundamental Research Funds for the Central Universities (Grant No ZYGX2016J081).

Appendix

1 Proof of Theorem 2

Proof If an adversary $\mathcal{F}$ can break the strong EUF-ID-CMA of the signature component of $\mathcal{IBCSESC}$, then one can construct an adversary $\mathcal{F}'$ that can break EUF-ID-CMA of the signature component of $\mathcal{IBCSE}$.

- **Initial**: $\mathcal{F}'$ is given the system public parameters *par* that it passes to $\mathcal{F}$.
- **Attack**: $\mathcal{F}'$ answers $\mathcal{F}$'s queries below.
  - For a key extraction query on *ID*, $\mathcal{F}'$ submits *ID* to its key extraction oracle and returns the result to $\mathcal{F}$.
  - For a signature query on (*m*,*ID*), $\mathcal{F}'$ submits $(0||m,ID)$ to its signature oracle and passes the result to $\mathcal{F}$.
  - For a decryption query on (c,ID), $\mathcal{F}'$ submits (c,ID) to its decryption oracle to get m'. If $m' = \perp$, $\mathcal{F}'$ returns $\perp$. Otherwise, $\mathcal{F}'$ parses m' as $tag||m$. Then, $\mathcal{F}'$ sets $tag = 0$ and returns m to $\mathcal{F}$.
  - For a signcryption query on (m, ID_s, ID_r), $\mathcal{F}'$ first submits $(1||ID_r||m, ID_s)$ to its signature oracle and gets the signature σ. Then, $\mathcal{F}'$ sets $tag = 1||ID_s$ and computes $c = \mathcal{IBCSE}.\mathbf{Encrypt}(par, tag||m||\sigma, ID_r)$. Finally, $\mathcal{F}'$ returns c to $\mathcal{F}$.
  - For an unsigncryption query on (c, ID_s, ID_r), $\mathcal{F}'$ submits (c, ID_r) to its decryption oracle to get m'. If $m' = \perp$, $\mathcal{F}'$ returns $\perp$. Otherwise, $\mathcal{F}'$ parses m' as $tag||m||\sigma$ and $tag = 1||ID_s$. $\mathcal{F}'$ checks if

$$\top = \mathcal{IBCSE}.\mathbf{Verify}(par, \sigma, 1||ID_r||m, ID_s)$$

    holds. If yes, $\mathcal{F}'$ returns m to $\mathcal{F}$. Otherwise, it returns $\perp$ to $\mathcal{F}$.

- **Forgery**: $\mathcal{F}$ finally outputs a message m^*, an identity ID^*, and a signature σ^*. If $\mathcal{F}$'s forgery is valid, then we have $\top = \mathcal{IBCSE}.\mathbf{Verify}(par, \sigma^*, 0||m^*, ID^*)$. In addition, $\mathcal{F}$ has not asked a key extraction query on ID^* and a signature query on (m^*, ID^*). Therefore, $\mathcal{F}'$ has not queried ID^* to its key extraction oracle and $(0||m^*, ID^*)$ to its signature oracle. $\mathcal{F}'$ outputs a message $0||m^*$, an identity ID^*, and a signature σ^*. It is obvious that $\mathcal{F}'$ succeeds with the same probability as $\mathcal{F}$.

Note that $\mathcal{F}$ can make a signcryption query on (m^*, ID^*, ID_r). However, the result is not useful since the returned $\sigma = \mathcal{IBCSE}.\mathbf{Sign}(par, 1||ID_r||m^*, ID^*, S_{ID^*})$. The signature verification equation $\mathcal{IBCSE}.\mathbf{Verify}(par, \sigma, 0||m^*, ID^*)$ must output the $\perp$ symbol. □

2 Proof of Theorem 3

Proof If an adversary $\mathcal{A}$ can break strong IND-ID-CCA2 of the encryption component of $\mathcal{IBCSESC}$, then one can construct an adversary $\mathcal{A}'$ that can break IND-ID-CCA2 of the encryption component of $\mathcal{IBCSE}$.

- **Initial**: $\mathcal{A}'$ is given system public parameters par that it passes to $\mathcal{A}$.
- **Phase 1**: $\mathcal{A}'$ answers $\mathcal{A}$'s queries as the proof of Theorem 2.
- **Challenge**: $\mathcal{A}$ generates two challenged messages (m_0, m_1) with the same length and an identity ID^*. ID^* has not been submitted to a key extraction query in Phase 1. So, $\mathcal{A}'$ has not queried ID^* to its key extraction oracle. $\mathcal{A}'$ sets $tag = 0$ and submits two messages $tag||m_0$, $tag||m_1$ and identity ID^* as its challenge. Finally, $\mathcal{A}'$ gets the challenged ciphertext c^* which it passes to $\mathcal{A}$.
- **Phase 2**: $\mathcal{A}$ can adaptively ask a polynomially bounded number of queries again as in Phase 1. $\mathcal{A}'$ answers $\mathcal{A}$'s queries as Phase 1. Since $\mathcal{A}$ cannot ask a key extraction query on ID^* and cannot ask a decryption query on (c^*, ID^*), $\mathcal{A}'$ does not submit ID^* to its key extraction oracle and does not submit (c^*, ID^*) to its decryption oracle.
- **Guess**: $\mathcal{A}$ finally outputs a guess γ' which $\mathcal{A}'$ outputs the same guess. It is obvious that $\mathcal{A}'$ succeeds with the same probability as $\mathcal{A}$.

$\mathcal{A}$ can ask an unsigncryption query on $(c^*(\text{or variations of } c^*), ID_s, ID^*)$. However, the result is not useful since we used different tags. In $\mathcal{IBCSESC}$.**Encrypt**, we encrypt $tag||m$ with $tag = 0$. In $\mathcal{IBCSESC}$.**Signcrypt**, we encrypt $tag||m||\sigma$ with $tag = 1||ID_s$. $\mathcal{A}$ must receive the $\perp$ symbol if it makes such a query. □

3 Proof of Theorem 4

Proof If an adversary $\mathcal{F}$ can break strong EUF-ID-CMA of the signcryption component of $\mathcal{IBCSESC}$, then one can construct an adversary $\mathcal{F}'$ that can break EUF-ID-CMA of the signature component of $\mathcal{IBCSE}$.

- **Initial**: $\mathcal{F}'$ is given system public parameters par which it passes to $\mathcal{F}$.
- **Attack**: $\mathcal{F}'$ answers $\mathcal{F}$'s queries as the proof of Theorem 2.
- **Forgery**: $\mathcal{F}$ produces a ciphertext c^*, a sender's identity ID_s^*, and a receiver's identity ID_r^*. $\mathcal{F}'$ first asks a key extraction query on ID_r^* to get $S_{ID_r^*}$. Then, $\mathcal{F}'$ computes $m' = \mathcal{IBCSE}.\textbf{Decrypt}(par, c^*, ID_r^*, S_{ID_r^*})$. If $\mathcal{F}$'s forgery is valid, then we have $m' = tag||m^*||\sigma^*$ and $tag = 1||ID_s^*$ such that $\top = \mathcal{IBCSE}.\textbf{Verify}(par, \sigma^*, 1||ID_r^*||m^*, ID_s^*)$. In addition, $\mathcal{F}$ has not asked a key extraction query on ID_s^* and a signcryption query on (m^*, ID_s^*, ID_r^*). So, $\mathcal{F}'$ has not queried ID_s^* to its key extraction oracle and $(1||ID_r^*||m^*, ID_s^*)$ to its signature oracle. $\mathcal{F}'$ outputs a message $1||ID_r^*||m^*$, an identity ID_s^*, and a signature σ^*. It is obvious that $\mathcal{F}'$ succeeds with the same probability as $\mathcal{F}$.

Note that $\mathcal{F}$ can ask a signature query on (m^*, ID_s^*). However, the result is not useful for $\mathcal{F}$ since the returned $\sigma = \mathcal{IBCSE}.\mathbf{Sign}(par, 0||m^*, ID_s^*, S_{ID_s^*})$. In unsigncryption algorithm, the verification $\mathcal{IBCSE}.\mathbf{Verify}(par, \sigma, 1||ID_r||m^*, ID_s^*)$ must output the $\perp$ symbol. □

4 Proof of Theorem 5

Proof If an adversary $\mathcal{A}$ can break the strong IND-ID-CCA2 of the signcryption component of $\mathcal{IBCSESC}$, then one can construct an adversary $\mathcal{A}'$ that can break IND-ID-CCA2 of the encryption component of $\mathcal{IBCSE}$.

- **Initial**: $\mathcal{A}'$ is given system public parameters par which it passes to $\mathcal{A}$.
- **Phase 1**: $\mathcal{A}'$ answers $\mathcal{A}$'s queries as the proof of Theorem 2.
- **Challenge**: $\mathcal{A}$ generates two challenged plaintexts (m_0, m_1) with the same length, a sender's identity ID_s^*, and a receiver's identity ID_r^*. ID_r^* has not been submitted to a key extraction query in Phase 1. So, $\mathcal{A}'$ has not queried ID_r^* to its key extraction oracle. $\mathcal{A}'$ first submits $1||ID_r^*||m_0$ and $1||ID_r^*||m_1$ to its signature oracle and gets signature σ_0^* and σ_1^*, respectively. Then, $\mathcal{A}'$ sets $tag = 1||ID_s^*$ and submits two messages $tag||m_0||\sigma_0^*$, $tag||m_1||\sigma_1^*$ and identity ID_r^* as its challenge. Finally, $\mathcal{A}'$ gets the challenged ciphertext c^* which it passes to $\mathcal{A}$.
- **Phase 2**: $\mathcal{A}$ can adaptively ask a polynomially bounded number of queries again as in Phase 1. $\mathcal{A}'$ answers $\mathcal{A}$'s queries as Phase 1. Since $\mathcal{A}$ cannot ask a key extraction query on ID_r^* and cannot ask an unsigncryption query on (c^*, ID_s^*, ID_r^*), $\mathcal{A}'$ does not submit ID_r^* to its key extraction oracle and does not submit (c^*, ID_r^*) to its decryption oracle.
- **Guess**: $\mathcal{A}$ finally outputs a guess γ' which $\mathcal{A}'$ outputs the same guess. It is obvious that $\mathcal{A}'$ succeeds with the same probability as $\mathcal{A}$.

Note that $\mathcal{A}$ cannot make an decryption query on $(c^*(\text{or variations of } c^*), ID_r^*)$. Similarly, the result is not useful since we use different tags in $\mathcal{IBCSESC}$.**Encrypt** and $\mathcal{IBCSESC}$.**Signcrypt**. $\mathcal{A}$ must receive the $\perp$ symbol if it makes such a query. □

References

1. Zheng, Y.: Digital signcryption or how to achieve cost (signature & encryption) ≪ cost (signature) + cost(encryption). In: Kaliski Jr., B.S. (ed.) CRYPTO'97. LNCS, vol. 1294, pp. 165–179. Springer, Heidelberg (1997)
2. An, J.H., Dodis, Y., Rabin, T.: On the security of joint signature and encryption. In: Knudsen, L.R. (ed.) EUROCRYPT 2002. LNCS, vol. 2332, pp. 83–107. Springer, Heidelberg (2002)
3. Baek, J., Steinfeld, R., Zheng, Y.: Formal proofs for the security of signcryption. J. Cryptol. **20**(2), 203–235 (2007)
4. Shamir, A.: Identity-based cryptosystems and signature schemes. In: Blakley, G.R., Chaum, D. (eds.) CRYPTO'84. LNCS, vol. 196, pp. 47–53. Springer, Heidelberg (1984)

5. Hess, F.: Efficient identity based signature schemes based on pairings. In: Nyberg, K., Heys, H. (eds.) SAC 2002. LNCS, vol. 2595, pp. 310–324. Springer, Heidelberg (2003)
6. Cha, J.C., Cheon, J.H.: An identity-based signature from gap Diffie-Hellman groups. In: Desmedt, Y.G. (ed.) PKC 2003. LNCS, vol. 2567, pp. 18–30. Springer, Heidelberg (2003)
7. Paterson, K.G., Schuldt, J.C.N.: Efficient identity-based signatures secure in the standard model. In: Batten, L., Safavi-Naini, R. (eds.) ACISP 2006. LNCS, vol. 4058, pp. 207–222. Springer, Heidelberg (2006)
8. Boneh, D., Franklin, M.: Identity-based encryption from the weil pairing. In: Kilian, J. (ed.) CRYPTO 2001. LNCS, vol. 2139, pp. 213–229. Springer, Heidelberg (2001)
9. Waters, B.: Efficient identity-based encryption without random oracles. In: Cramer, R. (ed.) EUROCRYPT 2005. LNCS, vol. 3494, pp. 114–127. Springer, Heidelberg (2005)
10. Boyen, X.: Multipurpose identity-based signcryption: a swiss army knife for identity-based cryptography. In: Boneh, D. (ed.) CRYPTO 2003. LNCS, vol. 2729, pp. 383–399. Springer, Heidelberg (2003)
11. Chen, L., Malone-Lee, J.: Improved identity-based signcryption. In: Vaudenay, S. (ed.) PKC 2005. LNCS, vol. 3386, pp. 362–379. Springer, Heidelberg (2005)
12. Barreto, P.S.L.M., Libert, B., McCullagh, N., Quisquater, J.J.: Efficient and provably-secure identity-based signatures and signcryption from bilinear maps. In: Roy, B. (ed.) ASIACRYPT 2005. LNCS, vol. 3788, pp. 515–532. Springer, Heidelberg (2005)
13. Haber, S., Pinkas, B.: Securely combining public-key cryptosystems. In: 8th ACM Conference on Computer and Communications Security (CCS 2001), Philadelphia, Pennsylvania, USA, pp. 215–224 (2001)
14. EMV Specifications, Version 4.2, Books 1–4 (June 2008). http://www.emvco.com/
15. Fan, J., Zheng, Y., Tang, X.: A single key pair is adequate for the Zheng signcryption. In: Parampalli, U., Hawkes, P. (eds.) ACISP 2011. LNCS, vol. 6812, pp. 371–388. Springer, Heidelberg (2011)
16. Paterson, K.G., Schuldt, J.C.N., Stam, M., Thomson, S.: On the joint security of encryption and signature, revisited. In: Lee, D.H., Wang, X. (eds.) ASIACRYPT 2011. LNCS, vol. 7073, pp. 161–178. Springer, Heidelberg (2011)
17. Chen, C., Chen, J., Lim, H.W., Feng, D.G.: Combined public-key schemes: the case of ABE and ABS. In: Takagi, T., Wang, G., Qin, Z., Jiang, S., Yu, Y. (eds.) ProvSec 2012. LNCS, vol. 7496, pp. 53–69. Springer, Heidelberg (2012)
18. Degabriele, J.P., Lehmann, A., Paterson, K.G., Smart, N.P., Strefler, M.: On the joint security of encryption and signature in EMV. In: Dunkelman, O. (ed.) CT-RSA 2012. LNCS, vol. 7178, pp. 116–135. Springer, Heidelberg (2012)
19. Vasco, M.I.G., Hess, F., Steinwandt, R.: Combined schemes for signature and encryption: the public-key and the identity-based setting. Inf. Comput. **247**, 1–10 (2016)

Cluster-Based Energy-Efficient Secure Routing in Wireless Sensor Networks

Tanmoy Maitra, Subhabrata Barman and Debasis Giri

Abstract To get inexpensive resolution of real-world problems such as weather forecasting, measurement of underground water label, traffic monitoring, activity of enemies, animals counting in forest, and so on, wireless sensor networks (WSNs) are widely used. Energy-efficient routing protocol is needed to provide the longevity of network lifetime by reducing power consumption of sensor nodes as well as whole networks. Besides, authenticity of sensor nodes and privacy of sensed data are needed in routing protocol for WSNs to provide secure communications, i.e., sensor-to-sensor as well as sensors-to-base station. Clustering technique provides an energy-efficient topology control approach. A minimum connected dominating set (MCDS) can be discovered by applying clustering technique which reduces power consumption in inter-cluster network routing. Cluster head and route selection can be used to provide an energy-efficient outer-cluster routing in WSNs. Identity-based scheme can be used to provide a secure secret message passing mechanism in WSNs. This paper has ng in wireless sensor network framework (E2SDRSNF) which consists of three proposed components. The proposed Algorithm 1 creates a MCDS which is used to build virtual backbone for energy-efficient inter-cluster routing. Data flows through the discovered virtual backbone nodes to base station via cluster heads which is discovered by applying proposed Algorithm 2. The proposed signcryption technique is used for secure communication in WSNs. The analysis of the proposed framework shows that it can save 2720 nJ/bit/m^2 energy than LEACH protocol for one communication (transmission/receiving) along with the security.

Keywords Attack · Clustering · Maximal independent set · Security · WSNs

---

T. Maitra (✉)
KIIT University, Bhubaneswar 751024, Odisha, India
e-mail: tanmoy.maitrafcs@kiit.ac.in

S. Barman · D. Giri
Haldia Institute of Technology, Haldia 721657, WB, India
e-mail: subhabrata_barman@yahoo.co.in

D. Giri
e-mail: debasis_giri@hotmail.com

P. Chandra et al. (eds.), *Information Technology and Applied Mathematics*, Advances in Intelligent Systems and Computing 699,
https://doi.org/10.1007/978-981-10-7590-2_2

1 Introduction

The wireless sensor networks (WSNs) consist of sensors which sense environmental states like pressure, temperature, motion, or pollutants from deployed area and pass those sensed data to base station via cluster head from. The key constraints of sensor are limited memory storage, low processing, and finite power supply. Maintenance cost is very high for deployed sensors. Replacement is the better choice than maintenance of deployed sensors. To increase network lifetime, many researchers pay attention on designing part of routing technique for WSNs. As there is no predefined infrastructure in WSNs, clustering is an effective topology control mechanism to surpass energy efficiency in routing protocols. Nowadays, to become routing protocols more energy-efficient, many researchers pay attention on cluster-based routing technique. Security in data communication is another salient issue while building routing protocol for WSNs. In clustering, every sensor node can be composed as an unit disk graph or general graph [1]. Sensor node can communicate via either single-hop or multi-hop communication [2]. Thus, it is needed to form a virtual backbone by constructing a MCDS [3] which is based on maximal independent set [4]. The virtual backbone increases efficiency of bandwidth and decreases communication overhead and overall energy consumption [5]. Therefore, lifetime of WSNs is enlarged.

Selected nodes have largest identity (power, degree, etc.), that produces very good dominating sets without knowing any positioning information of sensor nodes both for unit disk graph and non-unit disk graph. But, Gao et al. [6] show that “Largest ID” selected technique gives bad performance in worst case. Luby [4] proposed a fast distributed maximal independent set algorithm based on unit disk graph with known distance between two sensor nodes which produces a graceful and speedy solution to the problem of calculating a maximal independent set. Gao et al. [6] suggested that if sensor nodes are able to measure distances to their neighbors, then there is a chance to get a much faster solution to find minimum connected dominating set. Gao et al. [6] also proposed a possible algorithm for unit disk graph. Distributed greedy algorithm [7] provides distributed approximation algorithm using minimum dominating set for the arbitrary graphs. But, the time complexity of the algorithm, proposed in [7], gives bad performance for caterpillar graph [8]. Jia et al. [3] introduced an elegant distributed algorithm based on general graph which manages to overcome above-mentioned problem.

Heinzelman et al. [9] presented a protocol known as low energy adaptive cluster hierarchy (LEACH) to provide the maximum lifetime of network. Lindsey et al. [10] presented a power-efficient gathering in sensor information systems (PEGASIS) to provide better performance than LEACH algorithm. Kalpakis et al. [11] proposed a maximum lifetime data aggregation (MLDA) protocol based on an integer program to find an optimal flow network. Additionally, Kalpakis et al. built a data-gathering mechanism from the flow network in [11].

In cryptography, the main objective of encryption and digital signature is that no information about plaintext from ciphertext can be computed, and a valid signature on a message cannot be forged by any attacker, respectively. By taking advantages of aforementioned two cryptographic techniques, Zheng [12] first introduced the concept of signcryption. Latter, Boneh and Franklin [13] presented ID-based encryption scheme using Weil pairing. Chen and Lee [14] also demonstrated a signcryption scheme using bilinear pairing. Boldyreva et al. [15] presented a revocable IBE (RIBE) protocol to decrease the PKG's workload in Boneh and Franklin's scheme [13]. Selvi et al. [16] proposed a scheme for ID-based encryption and signcryption. Recently, Wu et al. [17] suggested a signature scheme and a group key exchange protocol, respectively. Lee et al. [18] also introduced a new revocable ID-based signcryption scheme.

This work proposes a framework referred as energy-efficient secure data routing for wireless sensor networks framework (E2SDRSN). There are three interacting components in E2SDRSN that are used to build a cluster-based energy-efficient secure routing:

1. Cluster formation using maximal independent set for unit disk graph (CFMI SUDG) to form network topology and also to find backbone nodes.
2. Cluster head and path selection technique (CPST) for providing energy-efficient routing.
3. Secure ID-based data routing (SIBDR) to provide secure communication among sensor nodes within the networks.

Organization of this study is as follows: Network and energy consumption models of the proposed framework are given in Sect. 2. Section 3 briefly describes some backgrounds which are used to construct the proposed framework. In Sect. 4, the proposed framework E2SDRSN is described. Analysis of E2SDRSN appears in Sect. 5. Finally, conclusion of this paper is provided in Sect. 6.

2 Network and Energy Consumption Models

There is a hierarchy among the nodes based on their potentiality in a hierarchical wireless sensor network (HWSN): base station (BS)/public key generator (PKG), cluster head (CH), and ordinary sensor node (SN). Network model is shown in Fig. 1.

Ordinary sensor node has finite battery power, small memory size, low data processing ability, and short range of radio transmission. The nodes can communicate with other sensor nodes within a cluster through cluster head (CH, i.e., also an ordinary sensor node). CH can communicate with other CHs directly and relay data between its cluster members and BS. Each CH in the network has unique identity. BS is a gateway to another network as well as an access point to the end users. BS has unlimited battery power, and it can compute complicated operations with a long transmission range. BS collects sensed data and performs costly operations to manage the network. It can be assumed that the base station is a trusted authority for the proposed framework. Here, the terms BS and private key generator (PKG) are used alternatively. Sensor nodes are placed around BS randomly. All the nodes are in the communication range of BS.

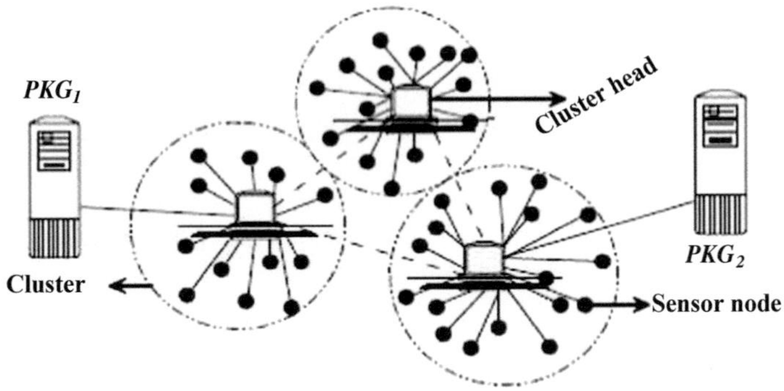

Fig. 1 Network model

This work divides the entire network into certain number of clusters where each cluster contains a CH. The CH communicates with other CH to deliver data to BS in multi-hop fashion. However, energy of a sensor node can be calculated as [19]:

$$E_{node} = bE_{sensN} + bE_{loggN} + bE_{elec} + bd_{toCH}^2 E_{amp} + E_{tranN} \quad (1)$$

where d_{toCH} is the distance between SN and CH, b denotes the number of bits in each packet, E_{amp} (=100 nJ/bit/m^2) is the free space fading energy, E_{sensN} is sensing activity of a node N, E_{loggN} is sensor logging energy, E_{elec} (=50 nJ/bit) is the energy dissipated to transmit/receive electronics, and E_{tranN} is total energy dissipation from the sensor node N per round. Here, E_{ele}, the energy loss, is d^2 for single-hop and d^4 for multi-hop communication.

Remaining energy $E_{re_i} = E_{ini_i} - E_{node_i}$, where $1 \leq i \leq n$; n is the total number of sensor nodes and threshold energy $E_T \geq \frac{\sum_{i=1}^{n} E_{node_i}}{n}$.

3 Preliminaries

This section briefly reviews the unit disk graph, connecting dominating set, cryptographic one-way hash function, and related mathematical problems.

Unit Disk Graph A graph $G = (V, E)$ is called as unit disk graph (UDG) [1]; if the Euclidean distance between $u \in V$ and $v \in V$ is less than or equal to 1, then there is an edge $e \in E$ between two nodes u and v. In the Euclidean plane, if $u = (x_1, y_2)$ and $v = (x_2, y_2)$, then the Euclidean distance is $d(u, v) = \sqrt{(x_1 - x_2)^2 + (y_1 - y_2)^2}$.

Connecting Dominating Set (CDS) In a graph $G = (V, E)$, a dominating set [3] is a subset $S \subseteq V$ such that every node is either in S or has at least one neighbor in S. In a graph $G = (V, E)$, a connected dominating set is a subset $S \subseteq V$ such that the subgraph induced by S is connected and S forms a dominating set in G.

Cryptographic One-Way Hash Function We refer articles [20–22].

Bilinear Pairing We refer article [13].

Bilinear Diffie–Hellman (BDH) Problem We refer article [13].

4 E2SDRSN: The Proposed Framework

This section introduces the proposed framework. Before going to deploy the nodes into the target area, base station (BS) performs the setup phase of the proposed SIBDR to generate the key pair for each node and burns the computed information into the memory of sensors as temper-resist. After deployment of nodes, the base station and nodes perform the proposed CFMISUDG to form the clusters and then perform the proposed CPST to find the efficient path for data transmission. During the data transmission, nodes and BS execute the proposed SIBDR in order to get secure data transmission. Table 1 shows the notations used in the proposed framework.

4.1 Cluster Formation Using Maximal Independent Set for Unit Disk Graph (CFMISUDG)

This section describes the topology control algorithm (see Algorithm 1). Initially, the radius r is initialized with the smallest value $2^{-\lambda}$, $\lambda \in \mathbb{N}$, and larger than d_{min} (a predefined static value). The radius r is increased by a factor of 2 in each iteration. For each iteration, all the nodes compute their $span\ d(v)$ (i.e. a node v be the number of uncovered nodes in v's neighborhood). Then it sends $(ID, E_{re_i}, span\ d(v))$ to other every nodes with distance 2. If any ID is missing, recalculate their span because there are some dead nodes in the network. A node v joins the set if it has the largest span in its double-hop neighbor nodes, $E_{rev} > E_T$ and $d_v < d_w$. Let N be the set of all nodes that are at most two hops far from v. The number of candidates that cover a node u be their support $s(u)$. The nodes form a maximal independent set $\mathcal{M}$ with probability $P(V)$. In each round of CFMISUDG, the node v computes the same operation until every neighbor nodes are covered.

4.2 Cluster Head and Path Selection Technique (CPST)

After deployment of sensor nodes, cluster heads and paths are selected once based on the three parameters: (1) weight of sensor nodes, (2) signal strength, and (3) path

Table 1 Nomenclature

Term	Usage	Term	Usage
E_{ini}	Initial energy of a node	$\mathcal{W}_v$	Weight of a node v
E_{node}	Energy consumption of a node	D_v	Average distance
n	Total number of nodes	P_v	Battery power of a node v
λ	Wavelength of signal	Δd_v	Degree of node v
$span\ d(v)$	Number of uncovered nodes in v's neighborhood	d	Average distance from node to neighbor nodes
$ID_{v/i}$	Identity of node v/i	P_{rx}	Receiving power of a signal
d_{min}	Minimum distance between two nodes	P_{tx}	Transmission power of a signal
M_v	Mobility of node v	SD_v	Distance between node and BS
N_{avg}	Average number of nodes within a cluster	$H(\cdot)$	One-way hash function
$\widehat{e}$	Bilinear mapping	(s, s')	Secret key of BS/PKG
P_{pub}	Public key of PKG	$ENC[r]_{P_{pub}}$	Encrypted r by public key P_{pub}
(X_i, B_i, A)	Stored in sensor chip	S_{pri_i}	Private key of node SN_i
S_{pub_i}	Public key of node SN_i	$\|\|$	Concatenation operation
$\oplus$	Xor operation	M	Plaintext/message
C	Ciphertext	σ_1	Signature
σ_2	Signcrypted text	G_1, G_2	Additive and multiplicative groups
P	Generator of G_1	q	(Prime) order of groups

table. If selected cluster heads will die in future, new cluster heads will be selected by running the proposed algorithm again.

In the proposed CPST (see Algorithm 2), BS chooses CH nodes and the optimal energy cost path from receiving signal strength and weight of every sensor nodes. CHs are selected with the highest weight and path is calculated as per receiving signal strength.

Algorithm 1 CFMISUDG

1: $r \leftarrow \min\{2^{-\lambda} | \lambda \in \mathbb{N} \wedge 2^{-\lambda} \geq d_{min}\}$
2: $v \leftarrow V$; $\mathcal{M} \leftarrow \Phi$;.
3: Each node calculates its *span* $d(v)$.
4: Each node sends $(E_{rev}, ID_v, span\ d(v))$ to all nodes within distance 2.
5: **if** any ID is missing **then**
6: Update its *span* and goto step 4.
7: **else**
8: Executes step 10.
9: **end if**
10: Calculates $\widehat{d}(v) \leftarrow d(v)$ rounded next smallest power of 2.
11: **while** $r \leq 1$ **do**
12: v is a candidate if $\widehat{d}(v) \geq \widehat{d}(w)\ \forall\ w \in N_v$, $E_{rev} > E_T$ and $dist(d_v, d_w) < r$
13: Computes cover $C(v)$ which is a set of uncovered nodes that v covers $|C(v)| = d(v)$.
14: Computes supports $s(u)$ that is number of candidates that cover u.
15: $s(u) \leftarrow |\{v \in V | u \in C(v)\}|$.
16: Each candidate joins Maximal Independent Set $\mathcal{M}$ with probability $P(V) \leftarrow 1/med(v)$, $med(v)$ = median support of all $u \in C(v)$.
17: $r \leftarrow r \cdot 2$.
18: **end while**

Algorithm 2 CPST

1: BS broadcast a message for CH selection.
2: All nodes send $(ID_v, P_v, D_v, M_v, \Delta d_v)$ to BS.
3: BS calculates $\mathcal{W}_v$ and P_{rx_v} for each node v.
4: BS puts the nodes in priority table by arranging in ascending order such that a node having highest weight and lowest P_{rx} is first in priority table and so on. If tie occurs, then BS puts the node in same row of priority table.
5: BS calculates $N_{avg} := \frac{N}{K}$, where N = Number of nodes and K = Number of clusters
6: $i := 1$.
7: **while** $i \neq N_{avg}$ **do**
8: BS takes the nodes from 5 consecutive rows of assigned priority table. If there are more than 1 nodes in a priority row, then BS takes a node which have highest weight.
9: **for all** selected node **do**
10: **if** $P_{rx_v} \leq P_{rx_w}$ **then**
11: Node v is selected and level i is assigned to it.
12: $i := i + 1$.
13: **else**
14: goto step 9
15: **end if**
16: **end for**
17: **end while**
18: BS creates a Path table based on level (see Table 2).
19: BS broadcast the Path table. All IDs' from path table are treated as cluster head.

4.2.1 Weight of Sensor Nodes

Weight of sensor nodes is depends on four parameters: (i) average of distance from node to neighbor node and distance between node to BS, (ii) battery power of the node, (iii) mobility of node, and (iv) degree of node. Weight of sensor node v is calculated as $\mathcal{W}_v = \{(w_1 \times D_v) + (w_2 \times P_v) + (w_3 \times \Delta d_v) + (w_4 \times M_v)\}$, where $\sum_{i=1}^{4} w_i = 1$ is the weighted value (here, 25% each to give equal priority).

We assume, if the Euclidean distance between two nodes is at most 1, then they are connected. Therefore, degree of nodes Δd_v is $\sum_{u,v \in V\ v \neq u}(dist(v,u) \leq 1)$. The running average of the speed for every node till current time T for mobility measurement can be computed as $M_v = \frac{1}{T}\sum_{t=1}^{T}[(x_t - x_{t-1})^2 + (y_t - y_{t-1})^2]^{\frac{1}{2}}$, where (x_t, y_t) and (x_{t-1}, y_{t-1}) are the coordinates of the node v at time t and $(t-1)$, respectively.

Power of node is measured based on the range of the broadcast message in particular area node. It is calculated as $P_{rx_v} = P_{tx_v} \times (\lambda/4\pi d)^2$.

Average distance D_v is the sum of average of the distance SD_v from node to neighbor node plus distance d between node to BS, i.e., $D_v = SD_v + d$, where $SD_v = (P_{tx_v}/P_{rx_v}) \times (\lambda/4\pi d)^2$.

4.2.2 Path Table

In the proposed algorithm (see Algorithm 2), path table is formed on the basis of assigning priority and level. The outcomes from path table are cluster heads, shortest and minimum energy cost routing path. Table 2 shows a path table where cluster head CH_2 will send data to BS through cluster head CH_1, and so on.

4.3 Secure ID-Based Data Routing (SIBDR)

In this section, we discuss proposed secure ID-based data routing scheme for providing secure communication among nodes and base station. The proposed scheme consists of setup phase, login phase, verification phase, key distribution phase, signcrypt phase, and unsigncrypt phase.

Table 2 Path table

ID	Level	Path
CH_1	1	–
CH_2	2	CH_2
CH_3	3	CH_2
⋮	⋮	⋮

4.3.1 Setup Phase

Base station or public key generator (PKG) computes the following operations:

1. PKG picks a security parameter $\mathbb{K} \in Z_n$ for some n and runs $\mathcal{G}$ on input $\mathbb{K}$ to produce a prime number q, two groups G_1 (additive), G_2 (multiplicative) of order $q > 2^{\mathbb{K}}$ and bilinear map $\widehat{e}$: $G_1 \times G_1 \rightarrow G_2$. PKG then picks a random generator $P \in G_1$.
2. PKG selects cryptographic hash functions H_1: $G_2 \times Z_q^* \rightarrow \{0,1\}^n$, H_2: $G_2 \rightarrow \{0,1\}^{2n}$, H_3: $Z_q^* \times Z_q^* \times \{0,1\}^n \rightarrow Z_q^*$, H_4: $G_1 \rightarrow Z_q^*$ and H_5: $\{0,1\}^n \times G_1 \times G_1 \times G_1 \times \{0,1\}^n \times \{0,1\}^n \times G_2 \rightarrow Z_q^*$ for some n. The security analysis of H_1, H_2, H_3, H_4, and H_5 will view as random oracles. The message space $M = \{0,1\}^n$ and a ciphertext space $C = \{0,1\}^{2n}$. Signature space is $\sigma_1 = G_1 \times \{0,1\}^n \times G_1 \times Z_q^* \times \{0,1\}^{2n}$ and signcryption space is $\sigma_2 = \{0,1\}^{2n} \times \{0,1\}^n \times G_1 \times G_1 \times G_1 \times Z_q^* \times Z_q^* \times Z_q^*$.
3. PKG picks random numbers $s, s' \in Z_q^*$. It computes $P_{pub} = [s]P$, $\alpha = \widehat{e}(P,P) \in G_2$, $\omega = H_1(\alpha, s)$ and $A = [s']P \in G_1$. For each node, PKG selects a random number $n_i \in Z_q^*$ and computes $X_i = (ID_i + \omega) \in \{0,1\}^n$, $S_{pri_i} = H_3(n_i, s', ID_i) \in Z_q^*$ and $B_i = [(s')^{-1} \cdot S_{pri_i}]P$. Then, PKG declares public parameters $(q, G_1, G_2, \widehat{e}, n, P, P_{pub}, H_1, H_2, H_3, H_4, H_5, \alpha, X_i)$ and keeps the secret parameters $(s, S^i_{pri}, s', \omega, ID_i)$ as a list. Finally, PKG stores (A, X_i, B_i) into the memory of sensor as tamper-resist.

4.3.2 Login Phase

PKG generates a broadcast message to verify every nodes in WSNs periodically. This broadcast message will be generated repeatedly after a certain period of time to get knowledge about the sensor nodes' activity. PKG maintains a broadcast message counter $count$. The value of $count$ is zero for the first broadcast message, and it will be incremented by one for each broadcast messages accordingly. After getting message from PKG, a sensor node SN_i (including cluster head) computes the following steps:

1. SN_i chooses $r_i, c_i \in Z_q^*$ randomly.
2. SN_i computes $W_i = [r_i]A$, $Z_i = [c_i]P + [r_i]B_i$, $L_i = Z_i + [(r_i)^{-1}]P$ and $S_{key_i} = ENC[r_i]_{P_{pub}}$. Then, it sends $\sigma_{i_1} = (W_i, X_i, L_i, c_i, S_{key_i})$ to cluster head CH_j.
3. The cluster head CH_j also computes its σ_{CH_j} (by following above two steps) and gathers the all incoming messages from sensor nodes by computing $\sigma_{total} = \langle\{||_{i=1}^{n_1}(\sigma_{i_1})\}, \sigma_{CH_j}\rangle$, where n_1 is the total number of sensor nodes within a cluster. Finally, cluster head CH_j sends σ_{total} to PKG through other cluster heads or directly according to the derived path table (see Table 2).

4.3.3 Verification Phase

After extracting signature σ_{i_1} $(=(W_i, X_i, L_i, c_i, S_{key_i}))$ from σ_{total}, PKG computes following operations to verify the sensor node SN_i:

1. Computes $ID_i = (X_i - \omega)$ and checks whether ID_i is present to its list. For the valid ID_i, PKG proceeds to next step, otherwise rejects SN_i.
2. Decrypts S_{key_i} to get r_i as $DEC[S_{key_i}]_s$ and checks $\widehat{e}(L_i, W_i) = \alpha^{(c_i\ r_i\ s' + r_i\ S_{key_i} + s')}$. If it holds, SN_i is authorized sensor; otherwise, PKG rejects SN_i.

$$\begin{aligned}
\text{Correctness: } \widehat{e}(L_i, W_i) &= \widehat{e}([c_i]P + [r_i]B_i + [(r_i)^{-1}]P, [r_i]A) \\
&= \widehat{e}([c_i]P, [r_i]A) \cdot \widehat{e}([r_i]B_i, [r_i]A) \cdot \widehat{e}([(r_i)^{-1}]P, [r_i]A) \\
&= \widehat{e}([c_i]P, A)^{r_i} \cdot \widehat{e}(B_i, A)^{r_i} \cdot \widehat{e}(P, A) \\
&= \widehat{e}(P, [s']P)^{c_i\ r_i} \cdot \widehat{e}([S_{pri_i}]P, P)^{r_i} \cdot \widehat{e}(P, [s']P) \\
&= \widehat{e}(P, P)^{c_i\ r_i\ s'} \cdot \widehat{e}(P, P)^{S_{pri_i}\ r_i} \cdot \widehat{e}(P, P)^{s'} \\
&= \alpha^{c_i\ r_i\ s'} \cdot \alpha^{S_{pri_i}\ r_i} \cdot \alpha^{s'} = \alpha^{(c_i\ r_i\ s' + r_i\ S_{pri_i} + s')}
\end{aligned}$$

4.3.4 Key Distribution

After verifying each sensor node SN_i, PKG provides private key and corresponding public key for all the nodes. PKG computes the following operations:

1. If $count = 0$, PKG computes $S_{pub_i} = [S_{pri_i}]P \in G_1$ and publishes S_{pub_i} as the public key of SN_i. PKG sends $EN_{key_i} = ENC[S_{pri_i}]_{r_i}$ to SN_i and increments $count$ by one. After getting EN_{key_i}, SN_i decrypts EN_{key_i} using key r_i to get S_{pri_i}. SN_i then checks $[S_{pri_i}]P$, and published S_{pub_i} are same or not. For the equality, SN_i stores the private key S_{pri_i} into its memory as temper-resist.
2. If $count > 0$, PKG chooses random number $RN_i \in Z_q^*$ and computes $S_{pri_i}^{new} = H_3(RN_i, s', ID_i) \in Z_q^*$, $B_i^{new} = [(s')^{-1} \cdot S_{pri_i}^{new}]P$, $S_{pub_i}^{new} = [S_{pri_i}^{new}]P \in G_1$ and publishes $S_{pub_i}^{new}$ as the public key of SN_i. Then PKG sends $EN_{key_i}^{new} = ENC[S_{pri_i}^{new} || B_i^{new}]_{r_i}$ to SN_i and increments $count$ by one. After getting $EN_{key_i}^{new}$, SN_i decrypts $EN_{key_i}^{new}$ using r_i to get $S_{pri_i}^{new}$ and B_i^{new}. SN_i checks $[S_{pri_i}^{new}]P$, and published $S_{pub_i}^{new}$ are same or not. For the equality, SN_i replaces B_i with B_i^{new} and stores the private key $S_{pri_i}^{new}$ into its memory as temper-resist.

4.3.5 Signcrypt Phase

This phase is invoked during the data communication. Suppose, a sensor node SN_f wants to send a data/message M to sensor node SN_i. SN_f has available information, i.e., $(M, X_f, X_i, S_{pub_i}, B_f, S_{pri_f})$. Then, SN_f performs the following steps to signcrypt the message M:

1. Chooses the random numbers $a, z, b \in Z_q^*$ and computes $C_1 = [a]P$, $C_2 = [b]P$, $C_3 = H_4(B_f) \in Z_q^*$, $C_4 = [(C_3 + z)b]P$, $C_5 = [C_3]C_2$, $C_6 = (S_{pri_f} \cdot h + a \cdot b) \in Z_q^*$, γ

$=\widehat{e}(S_{pub_i}, C_1)^a \in G_2, C = \{(M||X_f) \oplus H_2(\gamma)\} \in \{0,1\}^{2n}$ and $\mathcal{V} = (h \cdot a + z \cdot b)$, where $h = H_5(M, C_1, C_4, C_5, C_6, X_f, X_i, \gamma) \in Z_q^*$.

2. SN_f finally sends a signcrypt message $\sigma_f = (C, X_f, C_1, C_4, C_5, C_6, V, b)$ to SN_i.

4.3.6 Unsigncrypt Phase

After getting signcrypted cipher $\sigma_f = (C, X_f, C_1, C_4, C_5, C_6, \mathcal{V}, b)$ from SN_f, SN_i performs the following steps to unsigncrypt:

1. Computes $\gamma' = \widehat{e}(C_1, C_1)^{S_{pri_i}}$, $(M'||X'_f) = C \oplus H_2(\gamma')$ and checks X'_f, and received X_f are equal or not. For the inequality, SN_i rejects σ_f, otherwise goes to the next step.
2. Computes $h' = H_5(M', C_1, C_4, C_5, C_6, X_f, X_i, \gamma')$ and checks $\{[h']C_1 + C_4 - C_5\} = [\mathcal{V}]P$ and $\{[h']S_{pub_f} + [b]C_1\} = [C_6]P$. If both the conditions hold, SN_i accepts M' as original message M, otherwise rejects σ_f.

Correctness:

1. $\widehat{e}(S_{pub_i}, C_1)^a\ (=\gamma) = \widehat{e}([S_{pri_i}]P, C_1)^a = \widehat{e}([a]P, C_1)^{S_{pri_i}} = \widehat{e}(C_1, C_1)^{S_{pri_i}}$.
2. $\{[h']C_1 + C_4 - C_5\} = [h' \cdot a]P + [(C_3 + z)b]P - [C_3]C_2$
$= [h' \cdot a]P + [(C_3 + z)b]P - [C_3 \cdot b]P$
$= [h' \cdot a]P + [(C_3 + z - C_3)b]P$
$= [h' \cdot a]P + [z \cdot b]P$
$= [(h' \cdot a) + (z \cdot b)]P = [\mathcal{V}]P$ since, $h' = h$.
3. $\{[h']S_{pub_f} + [b]C_1\} = [h' \cdot S_{pri_f}]P + [b \cdot a]P = [(h' \cdot S_{pri_f}) + a \cdot b]P = [C_6]P$.

5 Performance Analysis of E2SDRSN

We use Sinalgo [23] simulator for testing and validating the proposed algorithms in E2SDRSN. Sinalgo is a java-based simulator for WSNs. Node distribution parameters for simulation are given in Table 3.

Connecting dominating set (CDS) size is proportional to the number of clusters in a networks. Number of clusters within a network is proportional to the energy consumption of networks as stated in [19]. Thus, small CDS size indicates the low power consumption in the networks. By these simulations, we can determine that the proposed CFMISUDG produces smaller CDS size than one round minimum dominating set algorithm as shown in Fig. 2. Thus, CFMISUDG is more energy-

Table 3 Node distribution setup in Sinalgo

Number of nodes	20–500	Mobility	Random direction
Distribution	Random	Reliability	Reliable delivery
Connectivity	Unit disk graph	Dimension (X)	dimX = 1000
Interference	No	Dimension (Y)	dimY = 1000

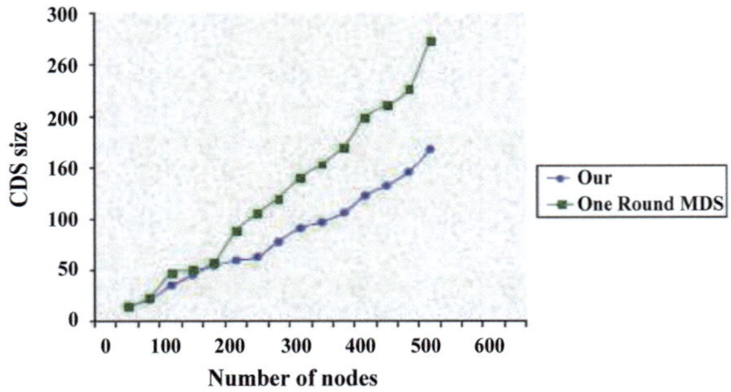

Fig. 2 CDS size comparison of proposed CFMISUDG with one round MDS algorithm for 20–500 nodes

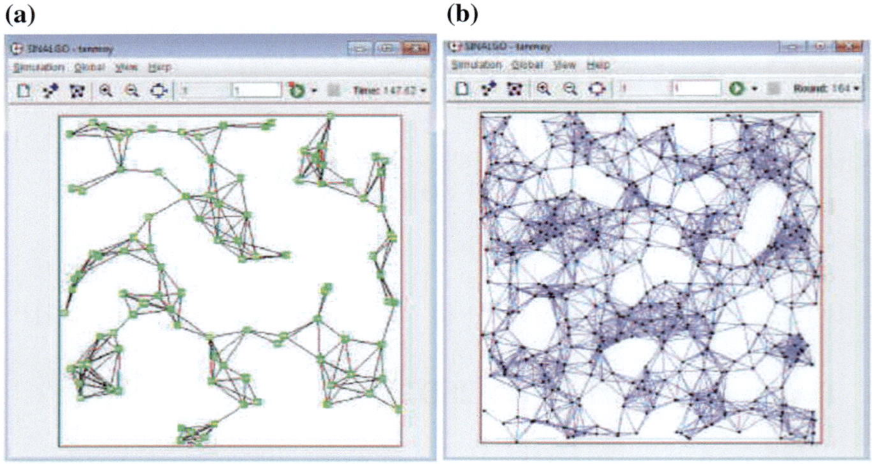

Fig. 3 CFMISUDG with **a** 100 nodes, and **b** 500 nodes

efficient than one round minimum dominating set algorithm. Figure 3 shows the snapshot of cluster formation using CFMISUDG.

Figure 4 depicts the energy consumed/bit by the proposed E2SDRSN, LEACH [9], and PEGASIS [10], respectively, with 20–200 nodes. From the simulation result, it can be said that for 20 nodes, LEACH protocol consumes energy = 20120 nJ/bit/m^2 and the proposed CFMISUDG consumes energy = 17400 nJ/bit/m^2 which saves 2720 nJ/bit/m^2 energy than LEACH protocol.

5.1 Security Analysis of SIBDR

Definition 1 *Confidentiality*: It can be said that the SIBDR scheme is semantically protected against an adaptive chosen ciphertext attack (SIBDR-CCA) if no proba-

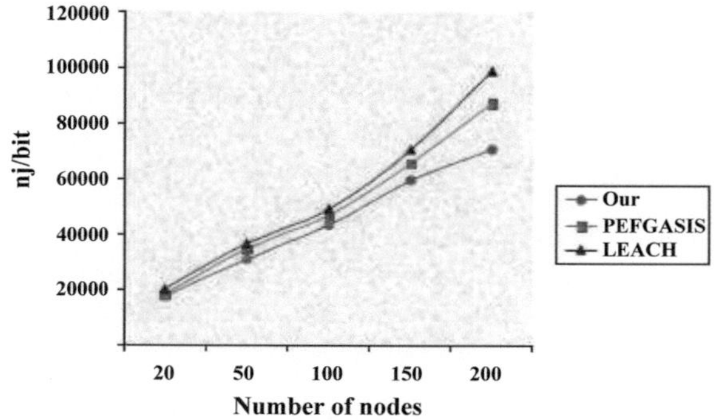

Fig. 4 Energy consumed/bit by E2SDRSN, LEACH [9], and PEGASIS [10]

bilistic polynomial-time attacker $\mathcal{A}$ has a non-negligible advantage in the following SIBDR-CCA game played between a challenger $\mathcal{C}$ and the attacker $\mathcal{A}$.

- **Setup**: The challenger $\mathcal{C}$ runs setup algorithm of SIBDR scheme to generate two master secret keys (s, s') and computes the public parameters $param$. Then, $\mathcal{C}$ computes $(S_{pri_i}, A, B_i, \omega)$ and sends $param$ to $\mathcal{A}$ and keeps tamper-resist parameters (B_i, A) (i.e., stored in sensor chip), s, s', and S_{pri_i} by itself.
- **Phase 1**: The attacker $\mathcal{A}$ may submit a number of several queries to the challenger $\mathcal{C}$ in an adaptive manner as follows:

- **Key extraction query**: There are two parts in this query: (1) login query and (2) key generation query.

1. *Login query*: After getting this query with stored tamper-resist parameters, $\mathcal{C}$ computes W_i, Z_i, L_i, S_{key_i} and sends signature $\sigma_{i_1} = (X_i, W_i, L_i, c_i, S_{key_i})$ to $\mathcal{A}$.
2. *Key generation query*: If verification procedure is successfully performed, $\mathcal{C}$ generates private and public keys for all verified sensor nodes and sends private key to that authenticated node and sends public key to $\mathcal{A}$; otherwise, it returns “Reject”.

- **Signcryption query**: Upon receiving this query on a message M, a sender's identity X_s, a receiver's identity X_r, receiver's public key S_{pub_r}, and sender's private key S_{pri_s}, $\mathcal{C}$ runs signcryption algorithm to generate a signcryptedtext σ_{r_2} and returns it to $\mathcal{A}$.
- **Unsigncryption query**: After getting this query for a signcryptedtext σ_{r_2}, $\mathcal{C}$ runs unsigncryption algorithm. $\mathcal{C}$ returns a message M, if σ_{r_2} is a correct signcryptedtext on M. Otherwise, it returns “Reject”.

$\mathcal{A}$ can submit his/her queries adaptively; that is, every request may depend on the response to the previous queries.

- **Challenge**: The adversary $\mathcal{A}$ outputs a sender's identity X_s^*, a receiver's identity X_r^*, receiver's public key $S^*_{pub_r}$, sender's private key $S^*_{pri_s}$, and a pair of distinct messages (M_0, M_1) to the challenger $\mathcal{C}$. $\mathcal{C}$ first randomly selects a value $\beta \in \{0, 1\}$. Then, $\mathcal{C}$ runs signcryption algorithm on $(M_\beta, X_s^*, X_r^*, S^*_{pub_i}, S^*_{pri_s})$ to return a signcryptedtext $\sigma^*_{r_2}$ and sends it to $\mathcal{A}$.

- **Phase 2**: The adversary $\mathcal{A}$ may make more queries which are the same as defined in Phase 1.
- **Guess**: If the adversary $\mathcal{A}$ produces his/her guess $\beta' \in \{0, 1\}$. Then, we can say that $\mathcal{A}$ wins SIBDR-CCA game if $\beta = \beta'$. Here, the advantage of $\mathcal{A}$ is defined as the probability.

Remark In the Definition 1, the adversary $\mathcal{A}$ is disallowed to make both login query on target X_s and unsigncryption query on target $(X_r, X_s, M, S_{pub_r}, S_{pri_s})$ in the SIBDR-CCA game because the target sensor's private key will be revealed and all information in sensor chip is temper-resist.

Theorem 1 *In the random oracle model, it is assumed that an attacker $\mathcal{A}$ with a non-negligible probability ϵ_1 can break ciphertext indistinguishability of SIBDR scheme under an adaptive chosen ciphertext attack. Then, there exists a challenger $\mathcal{C}$ with a non-negligible probability $\xi \geq 2\{\epsilon_1 - \epsilon(k)\} - \frac{q_{keyEx}+q_{Unsin}}{2^k} - \frac{q_{H_1}+q_{H_2}}{2^{2k}}$ which can solve the bilinear Diffie–Hellman (BDH) problem, where q_{H_1} and q_{H_2} queries to the random oracle H_1 and H_2, respectively, q_{keyEx} queries to key extraction oracle, and q_{Unsin} queries to unsigncryption oracle for any $0 \leq \epsilon(k) \leq \epsilon_1$.*

Proof It can be assumed that the challenger $\mathcal{C}$ gets a BDH instance $(P, [a]P, [b]P, [c]P)$ for $a, b, c \in_R Z_q^*$. After interacting with $\mathcal{A}$, $\mathcal{C}$ will provide the BDH solution $D = \widehat{e}(P, P)^{abc}$ in the SIBDR-CCA game by maintaining Definition 1.

- **Setup**: The challenger $\mathcal{C}$ executes setup algorithm to produce public parameters $params = \{\widehat{e}, G_1, G_2, q, P, P_{pub}, H_1, H_2, H_3, H_4, H_5, \alpha, X_i\}$ and sends it to $\mathcal{A}$, where P_{pub} is calculated as $[s]P$, $\alpha = \widehat{e}(P, P)$ and $X_i = \{ID_i + \omega\}$, where $\omega = H_1(\alpha, s)$. $\mathcal{C}$ then generates some public and private information for each sensor node i, i.e., $(S_{pri_i}, A, B_i, \omega)$, where $S_{pri_i} = H_3(n_i, s', ID_i)$, $B_i = [(s')^{-1}S_{pri_i}]P$, $A = [s']P$. The secret information $(s, s', S_{pri_i}, \omega)$ is kept by $\mathcal{C}$, and (A, B_i, X_i) is stored into the sensor chip as temper-resist. Note that, five hash functions behave as oracles controlled by the challenger $\mathcal{C}$.
- H_1 **query**: At any time, the attacker $\mathcal{A}$ can query the oracle H_1 with PKG secret key s and α. To answer the query, the challenger $\mathcal{C}$ maintains a list of tuples (α, s, ω, b_i) called L_{H_1}. If s has appeared in L_{H_1}, $\mathcal{C}$ returns ω $(=H_1(\alpha, s))$ to $\mathcal{A}$; otherwise, $\mathcal{C}$ selects $b_i \in_R \{0, 1\}^n$, responds to $\mathcal{A}$ with ω $(=b_i)$, and adds the tuple (α, s, b_i) into L_{H_1}. Note that, this list is initially empty.
- H_2 **query**: At any time, the attacker $\mathcal{A}$ can query the oracle H_2 with $\gamma \in_R G_2$. To answer this query, the challenger $\mathcal{C}$ maintains a list of tuples (γ, η, η_i) called L_{H_2}. If γ has appeared in L_{H_2}, $\mathcal{C}$ returns $H_2(\gamma)$ $(=\eta)$; otherwise, $\mathcal{C}$ randomly selects a string $\eta_i \in_R \{0, 1\}^{2n}$, responds to $\mathcal{A}$ with η_i, and adds the tuple (γ, η_i) into L_{H_2}. Note that, this list is initially empty.
- H_3 **query**: At any time, the attacker $\mathcal{A}$ can query the oracle H_3 with (n_i, s', ID_i). To answer this query, the challenger $\mathcal{C}$ maintains a list of tuples $(n_i, s', ID_i, c_i, S_{pri})$ called L_{H_3}. If (n_i, s', ID_i) have appeared in L_{H_3}, $\mathcal{C}$ returns $H_3(n_i, s', ID_i) = S_{pri}$; otherwise, $\mathcal{C}$ randomly selects $c_i \in_R Z_q^*$, responds to $\mathcal{A}$ with c_i $(=S_{pri})$, and adds the tuple (n_i, s', ID_i, c_i) into L_{H_3}. Note that, this list is initially empty.

- H_4 **query**: At any time, the attacker $\mathcal{A}$ can query the oracle H_4 with $B_f \in_R G_1$. To answer this query, the challenger $\mathcal{C}$ maintains a list of tuples (B_f, C_3, d_i) called L_{H_4}. If B_f has appeared in L_{H_4}, $\mathcal{C}$ returns $C3$ (=$H_4(B_f)$; otherwise, $\mathcal{C}$ selects $d_i \in_R Z_q^*$, responds to $\mathcal{A}$ with d_i (=C_3), and adds the tuple (B_f, d_i) into L_{H_4}. Note that, this list is initially empty.
- H_5 **query**: At any time, the attacker $\mathcal{A}$ can query the oracle H_5 with $(M, C_1, C_4, C_5, C_6, X_f, X_i, \gamma)$. To answer this query, the challenger $\mathcal{C}$ maintains a list of tuples $(M, C_1, C_4, C_5, C_6, X_f, X_i, \gamma, x_i, h_i)$ called L_{H_5}. If $(M, C_1, C_4, C_5, C_6, X_f, X_i, \gamma)$ has appeared in L_{H_5}, $\mathcal{C}$ returns $h = H_5(M, C_1, C_4, C_5, C_6, X_f, X_i, \gamma)$; otherwise, $\mathcal{C}$ randomly selects $x_i \in_R Z_q^*$, responds to $\mathcal{A}$ with x_i (=h), and adds the tuple (x_i, h) to the L_{H_5}. Note that, this list is initially empty.

– **Phase 1**: In this phase, the adversary $\mathcal{A}$ may build a number of several queries to the challenger $\mathcal{C}$ in an adaptive manner which are as follows:

- ***Key extraction query***: Firstly, upon receiving this query with PKG secret query s, $\mathcal{C}$ accesses the corresponding tuple (α, s, ω, b_i) from L_{H_1}. Then, $\mathcal{C}$ returns ω (=b_i) to the adversary $\mathcal{A}$. Then, $\mathcal{A}$ tries to extract ID_i by computing $(X_i - \omega)$ and proceed to H_3 query with (n_i, s', ID_i). Secondly, upon receiving this query, $\mathcal{C}$ accesses the corresponding tuple $(n_i, s', ID_i, c_i, Spri)$ from L_{H_3}. Then, $\mathcal{C}$ returns S_{pri} (=c_i) to adversary $\mathcal{A}$. After getting S_{pri} adversary $\mathcal{A}$ try to compute $B_i = [(s')^{-1}S_{pri}]P$.
- ***Signcryption query***: Upon receiving this query with $(M, X_s, X_r, S_{pub_r}, S_{pri_s})$, $\mathcal{C}$ accesses the corresponding tuple (B_f, C_3, d_i) from L_{H_4}, tuple (γ, η, η_i) from L_{H_2} and tuple $(M, C_1, C_4, C_5, C_6, X_f, X_i, \gamma, x_i, h_i)$ from L_{H_5}. Then, $\mathcal{C}$ randomly selects $a, z, b \in Z_q^*$ and computes $C_1 = [a]P$, $C_2 = [b]P$, $C_3 = H_4(B_f) \in Z_q^*$, $C_4 = [(C_3 + z)b]P$, $C_5 = [C_3]C_2$, $C_6 = (S_{pri_s} \cdot h + a \cdot b)$ and $\mathcal{V} = (h \cdot a + z \cdot b)$, where C_3 from L_{H_4} and h from L_{H_5}. $\mathcal{C}$ also computes $C = \{(M||X_f) \oplus H_2(\gamma)\}$, where γ (=$\widehat{e}(S_{pub}, C_1)^a$) from L_{H_2}. $\mathcal{C}$ sends $\sigma_2 = (C, X_s, C_1, C_4, C_5, C_6, \mathcal{V}, b)$ to the adversary $\mathcal{A}$.
- ***Unsigncryption query***: After getting this query with $(C, X_s, C_1, C_4, C_5, C_6, \mathcal{V}, b)$, $\mathcal{C}$ accesses the corresponding tuples (γ, η, η_i) from L_{H_2} and tuple $(M, C_1, C_4, C_5, C_6, X_f, X_i, \gamma, x_i, h_i)$ from L_{H_5}. Then, $\mathcal{C}$ picks γ from L_{H_2}, also picks h from L_{H_5}, and computes $(M||X_s) = C \oplus H_2(\gamma)$. Then, it checks $\{[h]C_1 + C_4 - C_5\} = [\mathcal{V}]P$ and $\{[h]S_{pub_f} + [b]C_1\} = [C_6]P$. If both conditions holds, then $\mathcal{C}$ returns M to $\mathcal{A}$; otherwise, returns ϕ.

Challenge: The adversary $\mathcal{A}$ outputs $(C^0, C^1, M_0, M_1, X_s, X_r, S_{pub_r}, S^*_{pri_s}, Spri_r{}^*, Spub_s)$ to the challenger $\mathcal{C}$. Then, $\mathcal{C}$ chooses a value $\beta \in_R \{0, 1\}$ and defines a signcryptedtext $\sigma_\beta = (C^\beta, X_s, C_1, C_4, C_5, C_6, \mathcal{V}, b)$ by computing $C_1 = [a]P$, $C_2 = [b]P$, $C_3 = d_i$, $C_4 = [(C_3 + z)b]P$, $C_5 = [C_3]C_2$, $C_6 = (S^*_{pri_s} \cdot x_i + a \cdot b)$, $\mathcal{V} = (x_i \cdot a + z \cdot b)$ and $C^\beta = \{(M_\beta||X_s) \oplus \eta_i\}$, where C_3 from H_4 query, x_i (=h) from H_5 query and η_i from H_2 query. Finally, $\mathcal{C}$ sends σ_{beta} to the adversary $\mathcal{A}$. It is easy to see that to unsigncrypt σ_{beta}, it is needed to compute $C^\beta \oplus H_2(\widehat{e}(C_1, C_1)^{S^*_{prir}})$ which requires proper knowledge about private key $Spri_r$ of the receiver.

– **Phase 2**: The adversary $\mathcal{A}$ may make key extraction query, signcryption query, unsigncryption query which are the same as defined in Phase 1.

– **Guess**: The adversary $\mathcal{A}$ outputs its guess $\beta' \in_R \{0, 1\}$ unless $\mathcal{A}$ queries key extraction to random oracle H_1 and H_3, and unsigncryption queries to random oracle H_2, H_4, and H_5. If $\mathcal{A}$ correctly guesses which C^β is for which M_β, then he/she wins the game.

By the assumption, there are two parts: one is hash queries to H_1 and H_3, and another part is hash queries to H_2, H_4, and H_5. $\mathcal{A}$ with a non-negligible probability ϵ_1 under an adaptive chosen ciphertext attack, we have to find the probability of $\mathcal{A}$ to get success in solving BDH. Let H_{q1} and H_{q2} denote an event that $\mathcal{A}$ asks the hash value of the tuple during the simulation of $\{H_1, H_3\}$, and $\{H_2, H_4, H_5\}$ accordingly. As we know that adversary wins the game if $\beta = \beta'$. Therefore, advantage over the game is defined as $ADV\ \mathcal{A}^{CCA} = 2Pr[\beta' = \beta] - 1$ where $pr[\beta' = \beta] = \frac{\epsilon_i+1}{2}$. We determine the probability for which $\beta' = \beta$. Firstly, we consider of first part: Let $evnt_1$ be an event in which attacker wins the game and $evnt_2$ be an event in which adversary $\mathcal{A}$ will not asked H_1 and H_3 hash queries. Then, probability of H_{q1} to happen is the same as in real attack. Thus, we have $Pr[evnt_1] = Pr[evnt_1|evnt_2]Pr[evnt_2] + Pr[evnt_1| \sim evnt_2]Pr[\sim evnt_2]$. It implies that:

$$\begin{aligned}
Pr[\beta' = \beta] &= Pr[\beta' = \beta| \sim H_{q1}]Pr[\sim H_{q1}] + Pr[\beta' = \beta|H_{q1}]Pr[H_{q1}] \\
&\leq Pr[\beta' = \beta| \sim H_{q1}](1 - Pr[H_{q1}]) + Pr[H_{q1}] \\
&\leq \frac{1+\epsilon(k)}{2}(1 - Pr[H_{q1}]) + Pr[H_{q1}] \\
\frac{\epsilon_i + 1}{2} &\leq \frac{1+\epsilon(k)}{2} + \frac{1}{2}Pr[H_{q1}]
\end{aligned}$$

Here, $\epsilon(k)$ is the maximal advantage of any passive attacker against the semantic security of key extraction scheme. Therefore, $Pr[H_{q1}] \geq \epsilon_1 - \epsilon(k)$. In case of simulation failure, one needs to handle the only two possibilities: in the first case, when the random challenge key extraction has been simulated to the key extraction oracle before the challenge phase, and in the second case, when the simulation is not correctly executed. The first event will occur with probability smaller than $\frac{q_{keyEx}}{2^K}$. In the second event, there is a chance in which key extraction oracle may reject the valid H_1 and H_3 queries. Since hash function H_1 is used to decide the acceptance of the PKG's secret key, the probability to reject the valid PKG's secret key is not greater than $\frac{q_{H_1}}{2^{2K}}$. Thus, the success probability ξ_1 of $\mathcal{A}$ is:

$$Pr[H_{q1}] - \frac{q_{keyEx}}{2^K} - \frac{q_{H_1}}{2^{2K}} \geq \epsilon_1 - \epsilon(k) - \frac{q_{keyEx}}{2^K} - \frac{q_{H_1}}{2^{2K}}$$

Hence, adversary $\mathcal{A}$ will be able to solve BDH with success probability $\xi_1 \geq \epsilon_1 - \epsilon(k) - \frac{q_{keyEx}}{2^K} - \frac{q_{H_1}}{2^{2K}}$. In case of second part, it will be happened same as previous for the unsigncryption queries. But, there H_2 is used to take decision on the acceptance of message. Thus, the probability to reject the ciphertext is not greater than $\frac{q_{H_2}}{2^{2K}}$.

Table 4 Comparison of computational cost of SIBDR with related schemes

Phase $\Rightarrow$	Signcryption			Unsigncryption		
Scheme $\Downarrow$	BP	SPM	HF	BP	SPM	HF
Wu et al. [17]	1	3	2	1	1	2
Lee et al. [18]	1	3	1	1	1	1
SIBDR	1	4	2	1	4	1

BP Bilinear pairing, *HF* Hash function, *SPM* Scalar point multiplication

Hence, adversary $\mathcal{A}$ will be able to solve BDH with success probability $\epsilon_2 \geq \epsilon_1 - \epsilon(k) - \frac{q_{Unsin}}{2^K} - \frac{q_{H_2}}{2^{2K}}$. So, total probability to solve BDH in SIBDR is ξ $(=\xi_1 + \xi_2) \geq 2\{\epsilon_1 - \epsilon(k)\} - \frac{q_{Unsin}+q_{keyEx}}{2^K} - \frac{q_{H_1}+q_{H_2}}{2^{2K}}$.

5.2 *Computation Analysis of SIBDR*

In Table 4, we show the comparison of computational cost of SIBDR with relative schemes for the signcryption and unsigncryption phases. SIBDR has high computational cost, but it is secure against SIBDR-CCA.

6 Conclusion

This work proposes a framework, called as energy-efficient secure data routing for wireless sensor network framework (E2SDRSN) which consists of three interacting components to build an energy-efficient secure routing protocol. Each of these algorithms reduces energy consumption up to certain level. E2SDRSN considers the communication as well as the computation limitations of sensor networks. As we have shown, the composition of the proposed algorithms can save 2720 nJ/bit/m^2 energy than LEACH protocol for one communication (transmission /receiving) and creation of cluster head and path selection within a cluster for 20 sensor nodes and 5 clusters in the WSNs. Future work will focus on the minimizing the cost factor within a cluster using any optimizing algorithm. Furthermore, we will reduce the computational cost of SIBDR in future.

References

1. Bagchi, A.: Sparse power-efficient topologies for wireless ad hoc sensor networks. In: IEEE International Symposium on Parallel Distributed Processing (IPDPS), pp. 1–10, April 2010
2. Maitra, T., Roy, S.: A comparative study on popular MAC protocols for mixed wireless sensor networks: from implementation viewpoint. Comput. Sci. Rev. **22**(C), 107–134 (2016)

3. Jia, L., Rajaraman, R., Suel, T.: An efficient distributed algorithm for constructing small dominating sets. Distrib. Comput. **15**(4), 193–205 (2002)
4. Luby, M.: A simple parallel algorithm for the maximal independent set problem. In: Proceedings of the Seventeenth Annual ACM Symposium on Theory of Computing, pp. 1–10, New York, NY, USA (1985)
5. Parthiban, N., Rajasingh, I., Rajan, R.S.: Minimum connected dominating set for certain circulant networks. Procedia Comput. Sci. **57**(C), 587–591 (2015)
6. Gao, J., Guibas, L., Hershberger, J., Zhang, L., Zhu, A.: Discrete mobile centers. In: Proceedings of the Seventeenth Annual Symposium on Computational Geometry, SCG '01, pp. 188–196, New York, NY, USA (2001)
7. Guha, S., Khuller, S.: Approximation algorithms for connected dominating sets. Algorithmica **20**(4), 374–387 (1998)
8. Ortiz, C., Villanueva, M.: Maximal independent sets in caterpillar graphs. Discret. Appl. Math. **160**(3), 259–266 (2012)
9. Heinzelman, W.R., Chandrakasan, A., Balakrishnan, H.: Energy-efficient communication protocol for wireless microsensor networks. In: Proceedings of the 33rd Hawaii International Conference on System Sciences, Washington, DC, USA (2000)
10. Lindsey, S., Raghavendra, C.S.: PEGASIS: power-efficient gathering in sensor information systems. In: Proceedings of the IEEE Aerospace Conference, vol. 3 (2002)
11. Kalpakis, K., Dasgupta, K., Namjoshi, P.: Efficient algorithms for maximum lifetime data gathering and aggregation in wireless sensor networks. Comput. Netw. **42**(6), 697–716 (2003)
12. Zheng, Y.: Digital signcryption or how to achieve cost (signature & encryption) « cost(signature) + cost(encryption), pp. 165–179. Springer, Berlin, Heidelberg (1997)
13. Boneh, D., Franklin, M.: Identity-Based Encryption from the Weil Pairing, pp. 213–229. Springer, Berlin, Heidelberg (2001)
14. Chen, L., Lee, J.M.: Improved identity-based signcryption. In: Proceedings of the 8th International Conference on Theory and Practice in Public Key Cryptography, pp. 362–379. Springer (2005)
15. Boldyreva, A., Goyal, V., Kumar, V.: Identity-based encryption with efficient revocation. In: Proceedings of the 15th ACM Conference on Computer and Communications Security, pp. 417–426. ACM (2008)
16. Selvi, S.S.D., Vivek, S.S., Rangan, P.C.: Identity Based Online/Offline Encryption and Signcryption Schemes Revisited, pp. 111–127. Springer, Berlin, Heidelberg (2011)
17. Wu, T.-Y., Tsai, T.T., Tseng, Y.M.: Revocable ID-based signature scheme with batch verifications. In: 2012 Eighth International Conference on Intelligent Information Hiding and Multimedia Signal Processing, pp. 49–54 (2012)
18. Lee, W., Seo, J.W., Lee, P.J.: Identity-Based Signcryption from Identity-Based Cryptography, pp. 70–83. Springer, Berlin, Heidelberg (2012)
19. Halgamuge, M.N., Zukerman, M., Ramamohanarao, K., Vu, H.L.: An estimation of sensor energy consumption. Prog. Electromagnet. Res. B **12**, 259–295 (2009)
20. Maitra, T., Giri, D.: An efficient biometric and password-based remote user authentication using smart card for telecare medical information systems in multi-server environment. J. Med. Syst. **38**(12) (2014)
21. Giri, D., Sherratt, R.S., Maitra, T.: A novel and efficient session spanning biometric and password based three-factor authentication protocol for consumer usb mass storage devices. IEEE Trans. Consum. Electron. **62**(3), 283–291 (2016)
22. Amin, R., Maitra, T., Giri, D., Srivastava, P.D.: Cryptanalysis and improvement of an RSA based remote user authentication scheme using smart card. Wire. Pers. Commun. **96**(3), 4629–4659 (2017)
23. Developed by the Distributed Computing Group at eth Zurich. www.disco.ethz.ch/projects/sinalgo/

A Two-Stage Approach for Text and Non-text Separation from Handwritten Scientific Document Images

Showmik Bhowmik, Soumyadeep Kundu, Bikram Kumar De, Ram Sarkar and Mita Nasipuri

Abstract The presence of non-text components in the document image hinders the result of an optical character recognition (OCR)-based document analysis system. Thus, text and non-text separation has become an essential task in the domain of document image processing. To address this issue, in the present work, a simple two-stage method is developed to separate the text and the non-text components from the images of handwritten scientific documents. Before starting the actual process, connected components from the document pages are extracted. Then, in the first stage, some commonly occurred components are identified and separated out as graphics. In the second stage, remaining components are passed through feature extraction and subsequent classification processes. Evaluating the system on handwritten scientific document images, it is found that 87.16% components are classified correctly as text or non-text.

Keywords Text/non-text separation · Handwritten image · Scientific document Straight line/almost straight line · Two-stage approach

S. Bhowmik · S. Kundu (✉) · B. K. De · R. Sarkar · M. Nasipuri
Jadavpur University, Kolkata 700032, WB, India
e-mail: soumyadeep1497@gmail.com

S. Bhowmik
e-mail: showmik.cse@gmail.com

B. K. De
e-mail: bikramkumarde@gmail.com

R. Sarkar
e-mail: raamsarkar@gmail.com

M. Nasipuri
e-mail: mitanasipuri@gmail.com

P. Chandra et al. (eds.), *Information Technology and Applied Mathematics*,
Advances in Intelligent Systems and Computing 699,
https://doi.org/10.1007/978-981-10-7590-2_3

1 Introduction

Documents are one of the vital ways used for sharing knowledge. To preserve the documents, either printed or handwritten, significant amount of research is going on to develop an OCR-based document image analysis system which can effectively transfer the information present in a document into its digital form. Before OCRing the contents of the document image, the first and foremost task is to separate the non-text or graphics from the text part of the document. This is because in any document analysis system, text and non-text elements are processed in a completely different manner. As a result, text and non-text separation reveals itself as an important prerequisite in the domain of document image processing.

Printed documents are considered as structured documents, whereas handwritten documents are treated as unstructured documents. Text components residing in printed documents are uniform in size and shape, which makes the text and non-text separation task easier in comparison with the handwritten documents, where shape and size of the components vary extremely. To address this issue, a novel two-stage method is developed here to separate text and non-text components found in handwritten scientific documents.

2 Related Work

The methods available in the literature for the separation of text and non-text can broadly be classified into three major categories, namely zone classification-based methods [1–4], connected component classification-based method [5–8], and pixel classification-based methods [9, 10]. Methods belonging to the first category classify an entire region extracted from a given document image as text or non-text region. Generally, these categories of methods use texture-based features to classify a given region. For example, researchers in [1, 2] have used gray level co-occurrence matrix (GLCM)-based features to classify the extracted region as text or non-text, whereas in [3], researchers have used features like mean brightness, mean-square deviation, mean difference of the mean brightness of blocks, etc. Methods presented in [4, 11] have used run-length-based features for purpose of region classification.

In contrast to first category, methods belonging to the second category classify the extracted components as text or non-text components. These methods generally use shape-based features for component classification. One such method is reported in [5], where researchers have used stroke width, elongation, solidity, and moment-based features to classify an extracted component as text or non-text. In [6], researchers have developed a recursive filter based on the connected components present in a given document image to separate the non-text from the text components. For that purpose, along with height, width, density of the extracted components, they have also considered the number of neighboring components. However, few researchers have also used features other than shape-based features to classify the components. For example, in [7], Spiral Run Length Smearing Algorithm (SRLSA)-based feature descriptor is used for component classification, whereas, in [8], local binary pattern (LBP)-based features are used to classify the components.

Although a significant number of attempts have been made to address the issue of text/non-text separation for printed documents, the attempts to solve the problem for handwritten documents are very few [7, 8]. This has been the prime motivation behind the present work.

3 Present Work

Handwritten documents are produced to serve a wide range of purposes and depending on that a significant difference is observed in the nature of the components present in the documents. For example, regular handwritten documents like letters, class notes, casual writings have large number of texts while other handwritten documents like laboratory reports of various science and engineering topics, maps have more number of non-text components. In this work, the application area is restricted to such scientific documents. A careful observation on the database used in the present work divulges that a large number of components have straight line/almost straight line (SL/ASL) like shapes, which may be generally the parts of tables or some circuit diagrams. Due to that fact, in the current work, the following Hypothesis is made regarding the components of handwritten documents under consideration:

Hypothesis a single component, which is a SL/ASL, is either a graphic or a part of a graphic.

Based on the *Hypothesis*, in the present work, a two-stage method is developed to isolate the non-text components from the text ones in the said type of handwritten documents.

Initially, the document images are binarized [12] and then connected components are extracted using an 8-way connected component labeling (CCL) algorithm [13]. After that these components are filtered based on the Hypothesis; that is, the connected components which basically represent a SL/ASL (either vertical, horizontal, or diagonal) are identified and recognized as non-text or graphic components. From the rest, some simple features are extracted to represent them at the feature space. Finally, the classification is performed using random forest (RF) classifier which is chosen by comparing its performance in terms of accuracy with some other well-known classifiers, viz. Naïve Bayes, support vector machine (SVM), bagging, and multilayer perceptron (MLP). Figure 1 shows the diagrammatic representation of the entire method.

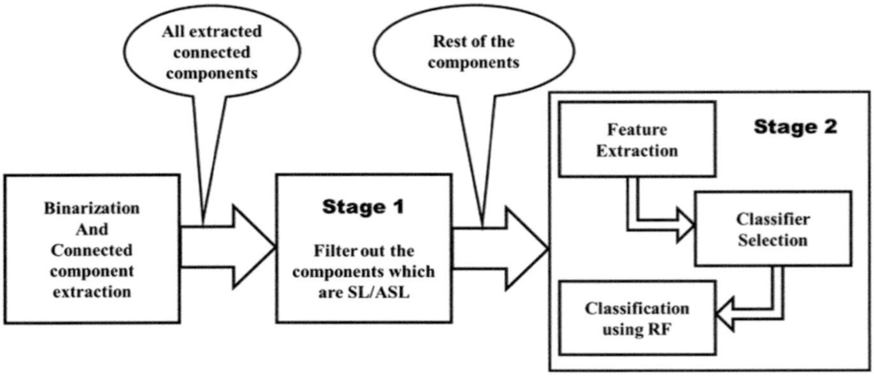

Fig. 1 Diagrammatic representation of the process followed to identify the text and non-text components

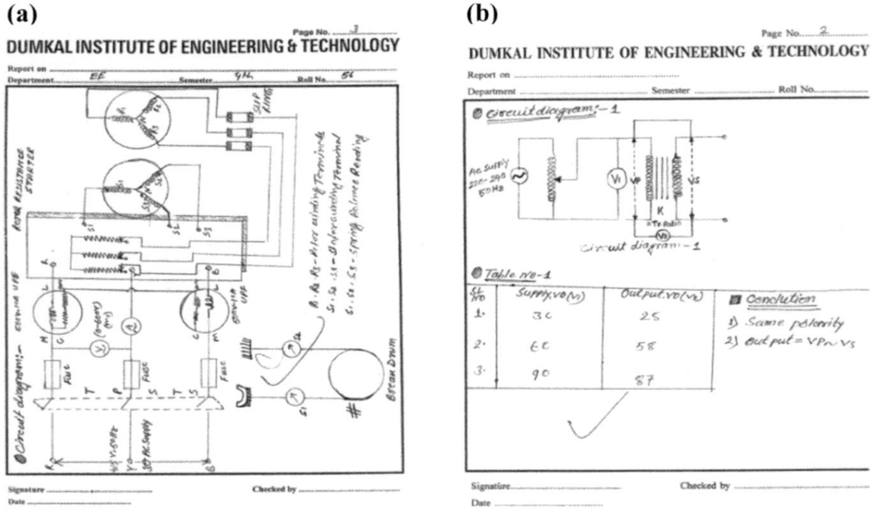

Fig. 2 Samples of engineering laboratory reports taken from the current database

3.1 Nature of the Database

In this work, 50 different, handwritten laboratory reports of different streams like Mechanical Engineering, Electrical Engineering, Physics, Chemistry are collected from various colleges. Out of those 50 pages, 35 pages also contain some printed text and the remaining pages are purely handwritten (sample images are shown in Fig. 2). All of these pages are digitized by a HP flatbed scanner at 300 dpi and saved as 24-bit BMP image format.

3.2 Image Binarization and Connected Component Extraction

All the images, used in the present work, are binarized using a method described in [12], which applies an adaptive foreground–background separation method on the input grayscale image to eliminate the background level noise and then uses Otsu's method [13] for final binarization. After binarization, an 8-way CCL algorithm [14] is used for component extraction. In this phase, initially the connected components are labeled using the said algorithm and then the minimum bounding box (MBB) for each labeled component is computed for further processing.

3.3 Stage 1: Separation and Recognition of Components Representing a SL/ASL

In the present work, some of the connected components are filtered out before sending them for feature extraction based on the said Hypothesis. This Hypothesis is set by observing the fact that in scientific documents, the shape and size of certain components are easily identifiable in comparison with components present in handwritten documents. As these documents mainly describe some technical experiments, they generally contain some tables, circuit designs, etc., where SL/ASL components are very common. Moreover, it can be said that these components are mostly non-text.

Keeping the above fact in mind, a module is developed in the present work that identifies the connected components representing SLs/ASLs. For that purpose, a given connected component is initially scanned columnwise to calculate the number of vertically isolated data pixels P_{ca} along with the number of data pixels having one or more vertical neighbors P_{cn}. If the value of P_{ca} is greater than P_{cn}, then it is considered as a SL/ASL-like component. Otherwise, the component is further scanned rowwise to compute the number of horizontally isolated data pixels P_{ra} and the number of data pixels having one or more horizontal neighbors P_{rn}. If P_{ra} is greater than P_{rn}, then also it is considered as a SL/ASL type component otherwise not. Figure 3 demonstrates the way that the pixels of a component are considered in the Stage 1 to check whether it represents a SL/ASL or not. In this figure, the pixels colored with *pink* represent the vertically isolated pixels. *Yellow*-colored pixels are both vertically and horizontally isolated. *Green*-colored pixels represent only horizontally isolated pixels, and the *gray*-colored pixels are neither horizontally nor vertically isolated. Now, in the first image of Fig. 3, the value of p_{ca} is 18 and P_{cn} is 10. As P_{ca} is greater than P_{cn}, so it is a component of type SL/ASL. On the other hand, in the second image, the value of p_{ca} is 6 and P_{cn} is 48. As P_{ca} is less than P_{cn}, the values of p_{ra} and P_{rn} need to be computed, which are 0 and 54, respectively. As again P_{ra} is less than P_{rn}, so it is not a SL/ASL component. The entire processes are summarized in Algorithm 1.

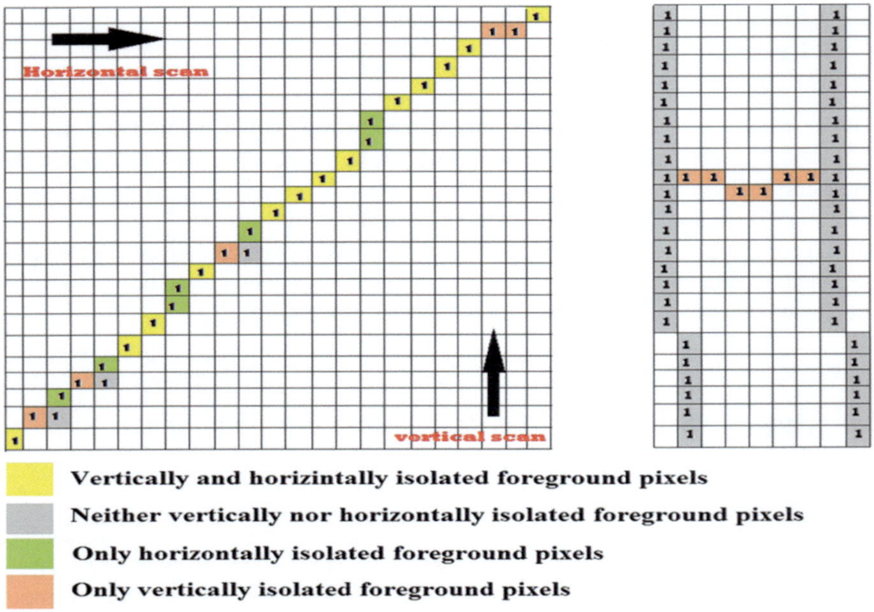

Fig. 3 Illustration of the process to identify a component as SL/ASL at Stage 1

ALGORITHM 1: *Identification of SL/ASL type components*

Input: Connected component
Output: SL/ASL or not

Perform thinning operation on the input connected component.
Scan the connected component column wise to find the followings
    P_{ca}, the number of foreground pixels which are residing isolated.
    P_{cn}, the number of foreground pixels which are attached with other foreground pixel(s) above or below it.
if $(p_{ca} > P_{cn})$
    the component is SL/ASL.
else
    Scan the connected component row wise to find the followings
        P_{ra}, the number of foreground pixels which are residing isolated.
        P_{rn}, the number of foreground pixels which are attached with other foreground pixel(s) before or after it.
    if $(p_{ra} > P_{rn})$
        the component is SL/ASL.
    else
        the component is not a SL/ASL
    end
end

Table 1 Performance comparison of MLP, SVM, Bagging, RF, and Naïve Bayes to select the final classifier for the classification at Stage 2

SL#	Classifier	Number of images	Number of components		Accuracy (in %)
			Text	Non-text	
1	MLP	20	3,226	2,814	86.81
2	SVM				86.73
3	Naïve Bayes				86.93
4	Bagging				85.56
5	RF				**88.23**

3.4 Stage 2: Feature Extraction and Classification

To reflect the effectiveness of the Hypothesis made at Stage 1, few simplistic features are extracted from those components which are passed to Stage 2. In other word, as it is supposed that most of the non-text components in the considered document type are of type SL/ASL; hence, a major portion of the separation is done at Stage 1. Thus, the components passed to Stage 2 do not need any complex features for classification. For that reason in this work, height and width of those components are taken as two feature values at Stage 2. Besides that, the ratio of the foreground pixel to the total pixel of the component (pixels inside the MBB of the components are considered) is taken as another feature value. Based on these feature values, all the extracted components at Stage 2 are classified as text or non-text. For the classification, a suitable classifier is also selected among five well-known classifiers based on their classification abilities. Details of the classifier selection and experimental results are described in Sect. 4.

4 Experimental Results and Error Analysis

In the present work, as mentioned earlier, a dataset of 50 images of handwritten laboratory reports are collected, from which 40,386 connected components are extracted. After filtering out 20,917 connected components at Stage 1, rest 19,469 components are passed through Stage 2 for feature extraction and subsequently for the classification.

For the classification of the components at Stage 2, a suitable classifier is chosen by comparing the performances of five well-known classifiers, viz. MLP, SVM, bagging, RF, and Naïve Bayes. These classifiers are executed on a small dataset of 20 handwritten reports selected randomly from the original dataset. Classifiers are set to default parameter values as given in WEKA [8]. Table 1 shows the accuracies achieved by those classifiers using threefold cross-validation scheme.

Table 2 Detail performance evaluation of the system on 50 images

Stage 1		Stage 2		Overall	
Number of components	Accuracy (%)	Number of components	Accuracy (%)	Number of Components	Accuracy (%)
20,917	84.52	19,469	90.21	40,386	87.16%

From Table 1, it is clear that RF performs better than other classifiers. Thus, RF is selected as the classifier for final classification at Stage 2.

The proposed system has achieved 84.52% and 90.21% accuracies at Stage 1 and Stage 2, respectively. Table 2 shows the overall performance of the system.

The overall accuracy O_A of the system is calculated by the following equation,

$$O_A = \left(\frac{Comp1}{Comp_{total}} \times acc_1\right) + \left(\frac{Comp2}{Comp_{total}} \times acc_2\right) \quad (1)$$

O_A	Overall accuracy achieved by the system.
$Comp1$	Number of components filtered out at Stage 1.
$Comp2$	Number of components passed to Stage 2.
$Comp_{total}$	Total number of components.
$acc1$	Accuracy achieved at Stage 1.
$acc2$	Accuracy achieved at Stage 2.

4.1 Error Analysis

Most of the components, either text or non-text, are identified correctly by the proposed system but there are some cases where the misclassifications are found. Figure 4a–c shows some misclassifications done by the system at different stages.

Here, the red color is used to represent text, whereas blue color represents the non-text; green-colored components are the components which are labeled as graphics at Stage 1. The black circles illustrate the misclassification occurred at Stage 1, and the red circles display the misclassification at Stage 2.

Though the *Hypothesis* made here is true for most of the non-text components, the same is also found true for some of the text components such as 'I' or 'i.' Besides that, in handwritten documents, during writing, multiple sub-strokes are created for a single character. These sub-components can appear as line-like components, which may cause misclassification. For example, four sub-components ('-,' '-,' '-,' and '|') can be created from 'E.' These are few obvious reasons of misclassification occurred at Stage 1.

In the present work, very simple features are used at Stage 2 to validate the effectiveness of the *Hypothesis* made at Stage 1. As a good number of non-text components are filtered out at Stage 1, the classification becomes less challenging

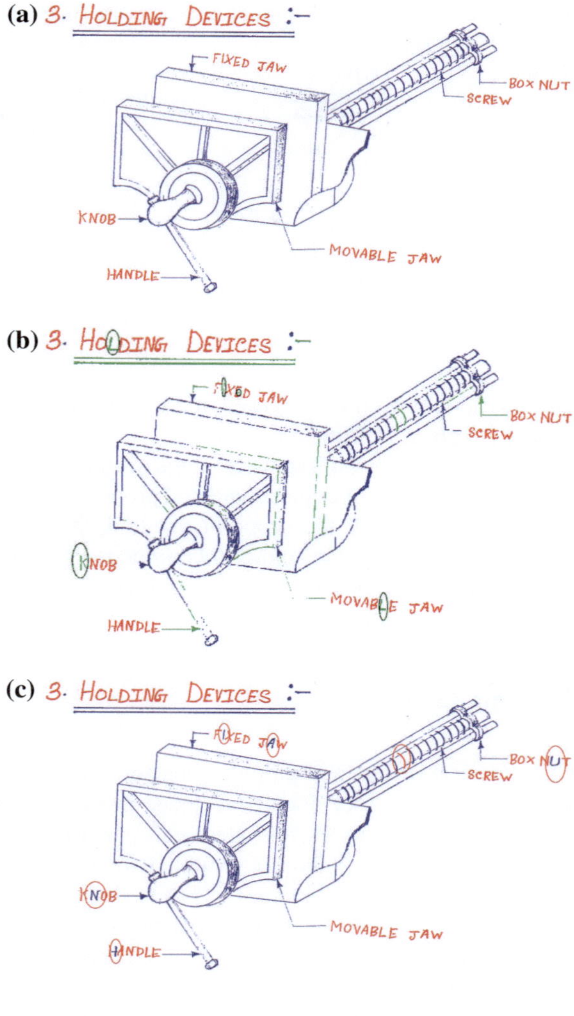

Fig. 4 **a** Sample image (texts are colored in red and non-texts are colored in blue), **b** sample misclassifications found at Stage 1 (green components are recognized as SL/ASL while black circles represent misclassifications), **c** sample misclassification at Stage 2 (red circles represent misclassification)

at Stage 2, which is the main reason of not using any computationally expensive features at Stage 2.

4.2 Performance Comparison with Some Existing Methods

In this subsection, the performance of the proposed work is compared with the performances of two existing methods introduced by Sarkar et al. [7] and Bhowmik et al. [8], respectively. During the comparison, the feature descriptors along with

Table 3 Performance comparison of the proposed method with two existing methods evaluated on the present dataset

Method	Feature descriptor	Size of the feature descriptor	Classifier	Accuracy (in %)
Sarkar et al. [7]	RLSA	7	SVM	80.78
Bhowmik et al. [8]	LBP	8	MLP	87.03
Proposed	SL/ASL, component height, width and pixel ratio	4	RF	**87.16**

the corresponding classifiers used in the said papers are considered. The comparison results are given in Table 3.

The comparison results show that the proposed method outperforms both of these existing methods. Although the method proposed by Bhowmik et al. [8] has achieved a result very close to the proposed method, they have used texture-based feature descriptor to classify the connected components which are computationally expensive than the features used in the current method.

5 Conclusion

Classification of text and graphics present in a document image is an unavoidable issue in the domain of document image processing. Without proper separation of text and non-text, no OCR system can perform well. Thus in the present work, a simple two-stage method is modeled for the isolation of non-text components from the text ones in the images of handwritten scientific documents. Though some misclassifications have been observed, the experimental results are quite satisfactory, which confirms the acceptability of the Hypothesis made earlier to filter out certain components.

To handle these misclassifications, a post-processing module can be added to the proposed system in future. Another plan is to evaluate the system on a relatively large set of document images to observe its scalability. Touching components are a common problem in handwritten image which also needs to be taken care of in future.

References

1. Oyedotun, O.K., Khashman, A.: Document segmentation using textural features summarization and feedforward neural network. Appl. Intell. 1–15 (2016)
2. Lin, M.W., Tapamo, J.-R., Ndovie, B.: A texture-based method for document segmentation and classification. South African Comput. J. **36**(1), 49–56 (2006)
3. Vil'kin, A.M., Safonov, I.V., Egorova, M.A.: Algorithm for segmentation of documents based on texture features. Pattern Recogn. Image Anal. **23**(1), 153–159 (2013)
4. Park, H.C., Ok, S.Y., Cho, H.: Word extraction in text/graphic mixed image using 3-dimensional graph model. ICCPOL **99**, 171–176 (1999)
5. Le, V.P., Nayef, N., Visani, M., Ogier, J.-M., De Tran, C.: Text and non-text segmentation based on connected component features in document analysis and recognition (ICDAR). In: 13th International Conference on 2015, pp. 1096–1100 (2015)
6. Tran, T.-A., Na, I.-S., Kim, S.-H.: Separation of text and non-text in document layout analysis using a recursive filter. KSII Trans. Inter. Inf. Syst. **9**(10), 4072–4091 (2015)
7. Sarkar, R., Moulik, S., Das, N., Basu, S., Nasipuri, M., Kundu, M.: Suppression of non-text components in handwritten document images. In: ICIIP 2011—Proceedings of International Conference Image Information Process on 2011, no. Iciip (2011)
8. Bhowmik, S., Sarkar, R., Nasipuri, M.: Text and non-text separation in handwritten document images using local binary pattern operator. In: Proceedings of the First International Conference on Intelligent Computing and Communication, 2017, pp. 507–515
9. Moll, M.A., Baird, H.S., An, C.: Truthing for pixel-accurate segmentation. In: Document Analysis Systems, 2008. DAS'08. The Eighth IAPR International Workshop on 2008, pp. 379–385
10. Moll, M.A., Baird, H.S.: Segmentation-based retrieval of document images from diverse collections. Electron. Imag. **2008**, 68150L–68150L (2008)
11. Shih, F.Y., Chen, S.S.: Adaptive document block segmentation and classification. IEEE Trans. Syst. Man Cybern. Part B **26**(5), 797–802 (1996)
12. Das, B., Bhowmik, S., Saha, A., Sarkar, R.: An adaptive foreground-background separation method for effective binarization of document images. In: International Conference on Soft Computing and Pattern Recognition pp. 515–524. Springer, Cham Dec. 2016
13. Otsu, N.: A threshold selection method from gray-level histograms. IEEE Trans. Syst. Man Cybern. **9**(1), 62–66 (1979)
14. AbuBaker, A., Qahwaji, R., Ipson, S., Saleh, M.: One scan connected component labeling technique. In: Signal Processing and Communications, 2007. IEEE International Conference on ICSPC 2007, pp. 1283–1286

The Approximate Solution for Multi-term the Fractional Order Initial Value Problem Using Collocation Method Based on Shifted Chebyshev Polynomials of the First Kind

Vijay Saw and Sushil Kumar

Abstract Nowadays, to survive and promote the market competition, multi-item business strategy is more effective for any production/manufacturing sector. Many physical problems can be best model by using fractional differential equation (FDE). In this paper, we propose the approximate scheme to solve multi-term fractional order initial value problem. The proposed scheme is based on collocation method and shifted Chebyshev polynomials (SCP). The fractional derivatives are utilized in the Caputo sense. The fractional order initial value problem can be reduced to a system of algebraic equations by utilizing the properties of SCP, which is solved numerically. The collocation point is chosen in such a way as to attain stability and convergence. The main theme of the proposal is to centralize the upper bound of the derived formula and convergence analysis. The numerical examples are achieved good accuracy using proposed scheme even by using small number of shifted Chebyshev polynomials.

Keywords Chebyshev polynomials · Collocation method · Caputo fractional derivative · Convergence analysis

1 Introduction

In recent years, fractional derivatives provide an excellent tool for the description of memory and hereditary properties of many materials. The fractional derivatives can be applied in turbulent flow, groundwater contaminant transport, chaotic of dynamics of classical conservative systems. The area of FDEs can be model successfully in several fields such as chemistry, biology, engineering, and physical science. For this reason, many authors are attracted in knowing the properties of fractional differential equations [1–3]. The primary idea of physical interest of FDE is more

V. Saw (✉) · S. Kumar
S.V. National Institute of Technology, Surat 395007, Gujarat, India
e-mail: vijaymaths89@gmail.com

S. Kumar
e-mail: skumar.iitr@gmail.com

P. Chandra et al. (eds.), *Information Technology and Applied Mathematics*,
Advances in Intelligent Systems and Computing 699,
https://doi.org/10.1007/978-981-10-7590-2_4

important. Mostly, we cannot find exact analytic solution of FDEs. However, numerical approximation scheme [4, 5] must be used for this purpose. Recently, various schemes have been investigated for the numerical solution of FDEs such as the homotopy-perturbation method [6], the variational iteration method [7], Adomian decomposition method [8, 9], and the homotopy analysis method [10, 11]. In [12], the authors used the predictor–corrector method to solve FDEs, whereas in [13], fractional order differential equation is solved using Caputo fractional derivative and takes advantage in terms of the fact that the FDEs is converted into Volterra integral equation. In [9], the authors approached for analytic and numerical solution for nonlinear FDEs. Moreover, the author in [14] gives the generalization of the Legendre operational matrix for solving FDEs, spectral method [15–19]. In [20], the author introduced spectral Tau collocation method with the help of Chebyshev polynomials to solve multi-term FDEs with the nonhomogeneous conditions, while Bhrawy et al. in [21] proposed to use quadrature Tau method to solve fractional order differential equations for variable coefficients. In [22], the author introduced numerical scheme for FDEs with two different delays.

The rest of the paper is as follows. In Sect. 2, we present the definition of Caputo fractional derivative and properties of Chebyshev polynomials. In Sect. 3, the approximate formula for this algorithm is given. In Sect. 4, the description of collocation method using Chebyshev polynomials for FDEs is given. In Sect. 5, some numerical examples are given and compared with other methods. Conclusion is presented in Sect. 6.

2 Preliminaries

2.1 Caputo Fractional Derivative

Definition 1 The fractional operator D^{μ}, $\mu > 0$, is defined in the Caputo sense [23]:

$$D^{\mu}\nu(x) = \frac{1}{\Gamma(m-\mu)}\int_{0}^{x}\frac{\nu^{(m)}(t)}{(x-t)^{(\mu-m+1)}}dt, m-1 < \mu \leq m, m \in N, x > 0. \quad (1)$$

The Caputo fractional derivative satisfies linearity properties:

$$D^{\mu}(\lambda p(x) + \sigma q(x)) = \lambda D^{\mu} p(x) + \sigma D^{\mu} q(x). \quad (2)$$

The Caputo fractional derivative in the power of x is defined as :

$$D^{\mu} K = 0, \quad \text{K is a constant}, \quad (3)$$

$$D^{\alpha} x^{\beta} = \begin{cases} \frac{\Gamma(\beta+1)}{\Gamma(\beta-\alpha+1)} x^{\beta-\alpha}, & \text{for } \beta \in N_0 \text{ and } \beta \geq \lceil \alpha \rceil \\ 0, & \text{for } \beta \in N_0 \text{ and } \beta < \lceil \alpha \rceil, \end{cases} \quad (4)$$

For more properties about the fractional derivatives, see [23, 24].

3 Chebyshev Polynomials of the First Kind

Definition 2 The Chebyshev polynomials of the first kind is defined by the relation [25]:

$$T_n(x) = \cos(n\theta),$$

where $x = \cos(\theta)$ and $\theta \in [0, \pi]$.

The polynomials $T_n(x)$ are orthogonal on $[-1, 1]$ with respect to the weight function $w_1(x)$:

$$\int_{-1}^{1} T_n(x)T_k(x)w_1(x)dx = \begin{cases} 0, & n \neq k, \\ \frac{\pi}{2}, & n = k, \ n \neq 0, \end{cases} \tag{5}$$

where $w_1(x) = \frac{1}{\sqrt{(1-x^2)}}$. The polynomials $T_n(x)$ are generated by the fundamental recurrence relations

$$T_{n+1}(x) = 2xT_n(x) - T_{n-1}(x), \quad n = 1, 2, \ldots,$$

with the conditions

$$T_0(x) = 1, \quad T_1(x) = x.$$

The polynomials $T_p(x)$ can be written in analytic form in the following form:

$$T_p(x) = p\sum_{k=0}^{\lceil \frac{p}{2} \rceil}(-1)^k 2^{p-2k-1}\frac{(p-k-1)!}{(k)!(p-2k)!}x^{p-2k}, \quad p > 0, \tag{6}$$

where $\lceil \frac{p}{2} \rceil$ indicates the integer part of $p/2$.

Sometimes, the range of the variable x in the $[0, 1]$ is more convenient for our purpose rather than the interval $[-1, 1]$, so we transform the independent variable $x \in [0, 1]$ to the variable t in $[-1, 1]$ by the substitution $t = 2x - 1$ or $x = \frac{1}{2}(1+t)$, and thus, $T_n^*(x)$ is called shifted Chebyshev polynomials on $[0, 1]$ by the relation

$$T_n^*(x) = T_n(2x - 1).$$

These polynomials are orthogonal on $[0, 1]$ with the weight function $w_2(x)$:

$$\int_0^1 T_n^*(x)T_s^*(x)w_2(x) = \begin{cases} 0, & n \neq s, \\ \pi, & n = s = 0, \\ \frac{\pi}{2}, & n = s \neq 0, \end{cases} \tag{7}$$

where $w_2(x) = (x - x^2)^{-\frac{1}{2}}$. The polynomials $T_n^*(x)$ are generated by the following fundamental recurrence relations

$$T_{n+1}^*(x) = 2(2x - 1)T_n^*(x) - T_{n-1}^*(x), \quad n = 1, 2, \ldots,$$

with the conditions

$$T_0^*(x) = 1, \quad T_1^*(x) = 2x - 1.$$

The polynomials $T_r^*(x)$ can be written in analytic form in the following form:

$$T_r^*(x) = r\sum_{s=0}^{r}(-1)^{r-s}\frac{2^{2s}(r + s - 1)!}{L^s(2s)!(r - s)!}x^s, \ r > 0. \tag{8}$$

The approximation function $g(x)$, which is measurable in space $[0, L]$, can be explicit in the form of $T_i^*(x)$ as follows:

$$g(x) = \sum_{i=0}^{\infty} c_i T_i^*(x), \tag{9}$$

where c_i are constants and first term of the series is halved. In particular, the approximation function $g(x)$ is approximated by the finite sum of $(m + 1)$-terms such as

$$g_m(x) = \sum_{i=0}^{m} c_i T_i^*(x), \tag{10}$$

where the first term of the series is halved and coefficients c_i are calculated in the following ways

$$c_0 = \frac{2}{\pi}\int_0^L \frac{g(x)T_0^*(x)}{\sqrt{Lx - x^2}}dx, \tag{11}$$

$$c_i = \frac{2}{\pi}\int_0^L \frac{g(x)T_i^*(x)}{\sqrt{Lx - x^2}}dx, \ i = 1, 2, \ldots, m. \tag{12}$$

4 Description of the Approximate Formula and Error Analysis

Theorem 1 *Let the approximation function* $g(x)$ *be approximated by Chebyshev polynomials which is defined in* (10) *and assume that* $\mu > 0$*, then*

$$D^{\mu}(g_m(x)) = \sum_{i=\lceil\mu\rceil}^{m} \sum_{k=\lceil\mu\rceil}^{i} c_i w_{i,k}^{(\mu)} x^{k-\mu}, \tag{13}$$

where $w_{i,k}^{(\mu)}$ *is given by:*

$$w_{i,k}^{(\mu)} = (-1)^{i-k} \frac{2^{2k} i (i+k-1)! \Gamma(k+1)}{L^k (i-k)! (2k)! \Gamma(k+1-\mu)}. \tag{14}$$

Proof The Caputo fractional derivative satisfies linear properties, and we have

$$D^{\mu}(g_m(x)) = \sum_{i=0}^{m} c_i D^{\mu}(T_i^*(x)). \tag{15}$$

Applying Eqs. (3) and (4), we get

$$D^{\mu}(T_i^*(x)) = 0, \quad i = 0, 1, \ldots, \lceil\mu\rceil - 1, \ \mu > 0. \tag{16}$$

Also, for $i = \lceil\mu\rceil, \lceil\mu\rceil + 1, \ldots, m$, and by using Eqs. (3) and (4), we get

$$\begin{aligned} D^{\mu} T_i^*(x) &= i \sum_{k=\lceil\mu\rceil}^{i} (-1)^{i-k} \frac{2^{2k}(i+k-1)!}{L^k (i-k)!(2k)!} D^{\mu} x^k \\ &= i \sum_{k=\lceil\mu\rceil}^{i} (-1)^{i-k} \frac{2^{2k}(i+k-1)! \Gamma(k+1)}{L^k (i-k)!(2k)! \Gamma(k+1-\mu)} x^{k-\mu}. \end{aligned} \tag{17}$$

Now combining the Eqs. (14), (16), and (17) goes to the desired results which shows the complete of the theorem.

Error Analysis

Theorem 2 (Chebyshev Truncation Theorem) *The approximation function* $g(x)$*, the sum of its first* $(m+1)$ *terms is absolutely bounded through the sum of the values of all the neglected coefficients. If*

$$g_m(x) = \sum_{i=0}^{m} c_i T_i^*(x) \tag{18}$$

then

$$E_T(m) \equiv\mid g(x) - g_m(x) \mid\leq \sum_{i=m+1}^{\infty} \mid c_i \mid, \ x \in [0, 1]. \tag{19}$$

Proof Since we know that the maximum value of shifted Chebyshev polynomials is one, that is $|T_i^*(x)| \leq 1$ for all $x \in [0, 1]$ and for all i. Therefore, the approximating function of the ith term is bounded by coefficients c_i and subtracting the $(m + 1)$-terms series from the infinite series, which gives the difference of each terms is bounded by the coefficients and summing the difference of bounding of each terms get the desired result.

Theorem 3 *If $g(x)$ is n-times (n) continuously differentiable in [−1, 1] and $g_m(x)$ is also continuously differentiable, then the rate of convergence of Chebyshev polynomial is*

$$c_i = O(m^{-n}), m = 1, 2, \ldots.$$

Proof By integrating Eqs. (11) and (12) and using the definition of integration of parts, we get

$$c_i = \frac{1}{m^n}\frac{2}{\pi}\int_{-1}^{1}(1-x^2)^{-\frac{1}{2}} f^n(x)T_m(x)dx.$$

Since the maximum value of $T_m(x)$ is 1 on $[-1, 1]$, it means that the truncation error for the continuous expansion is bounded by the sum of the absolute value of the neglecting coefficients

$$|g(x) - g_m(x)| \sum_{m=m+1}^{\infty} |c_i|.$$

This can also hold for the interpolating polynomial sum

$$|g(x) - I_m(x)|2 \sum_{m=m+1}^{\infty} |c_i|, \ x \in [-1, 1].$$

Combining the above, we get

$$|g(x) - g_m(x)| = O(m^{-n}), m = 1, 2, \ldots.$$

Note: If $g(x)$ is infinitely differentiable, then convergence is faster than $O(m^{-n})$, it does not mean how large value n, and it is known as exponential accuracy.

5 Collocation Method for Multi-term FDEs

It is known that the differential equations in physical space can be approximated using collocation method. We can easily apply in several problems such as variable coefficient and nonlinear differential equations. In this section, an algorithm for solving fractional order differential equation is proposed based on Chebyshev collocation method.

$$D^{\mu}v(x) = H\left(x, v(x), D^{\gamma_1}v(x), \ldots, D^{\gamma_k}v(x)\right), \; x \in (0, 1), \tag{20}$$

with initial conditions

$$v^{(i)}(0) = d_i, \; i = 0, 1, 2, \ldots, m-1, \tag{21}$$

here fractional operator D^{μ} is defined in the Caputo sense, and H is nonlinear function. For approximation, we use the shifted Chebyshev polynomials as a basis function and applied collocation method for solving multi-term FDE (20) with the initial conditions (21) has a unique continuous solution its satisfies Lipschitz criterion. Let

$$v_m(x) = \sum_{i=0}^{m} c_i T_m^*(x), \tag{22}$$

where c_i are unknown coefficients and first term of the series is halved. By virtue of Theorem 1, the fractional order derivatives $D^{\mu}v(x), D^{\gamma_1}v(x), \ldots, D^{\gamma_k}v(x)$ can be expressed explicitly in terms of the expansion coefficients c_i. For applying the Chebyshev collocation method, the criterion for solving Eq. (20) subject to initial conditions (21) is to find $v_m(x) \in S_m(0, L)$ in such a way

$$D^{\mu}v_m(x) = H\left(x, v_m(x), D^{\gamma_1}v_m(x), \ldots, D^{\gamma_k}v_m(x)\right) \tag{23}$$

and the fact that $v_m(x)$ must satisfy the fractional differential equation in some suitably chosen collocation points $\theta_i, \; i = 1, 2, \ldots, m$. The convergence of numerical solution and its computational stability gets affected by the particular choice of collocation points. In order to find the unknown coefficients, Chebyshev collocation method with collocation points

$$\theta_i = \frac{1}{2} + \frac{1}{2}\cos(i\frac{\pi}{m}), i = 1, 2, \ldots, m \tag{24}$$

$$\sum_{i=0}^{m} c_i D^{\mu}T^*(x_{\theta_i}) = H\left(x_{\theta_i}, \sum_{i=0}^{m} c_i D^{\mu}T^*(x_{\theta_i}), \sum_{i=0}^{m} c_i D^{\gamma_1}T^*(x_{\theta_i}), \ldots, \sum_{i=0}^{m} c_i D^{\gamma_k}T^*(x_{\theta_i})\right),$$
$$i = 1, 2, \ldots, m, \tag{25}$$

with (21) written in the form

$$\sum_{i=0}^{m} c_i T_i^*(0) = d_i, \quad i = 0, 1, 2, \ldots, m - 1. \tag{26}$$

For finding the unknown coefficients $c_i (i = 0, 1, 2, \ldots, m)$, combining the Eqs. (25)–(26), we get the system of $(m + 1)$ algebraic equations which is solved numerically.

Boundary Value Problem For the boundary value problem, the boundary conditions (for m is even) are defined as

$$v^{(i)}(0) = a_i, \ v^{(i)}(L) = b_i, \ i = 0, 1, \ldots, \frac{m}{2} - 1. \tag{27}$$

We can use the same technique as described in Sect. 4, for the FDE (20) but Eq. (21) shall be changed to be (27). In order to finding the unknown coefficients $c_i (i = 0, 1, 2, \ldots, m)$, combine the Eqs. (25)–(27), and we get the system of $(m + 1)$ linear algebraic equations which can be solved numerically by iteration method.

6 Numerical Experiments

In this section, the proposed scheme is implemented for solving multi-term fractional order IVP and comparing with existing method.

Example 1 Let us consider the fractional order IVP [26]

$$D^{(\frac{3}{2})} v(x) + 3v(x) = 3x^3 + \frac{8}{\sqrt{\pi}} x^{\frac{3}{2}}, \tag{28}$$

with the initial conditions

$$v(0) = 0, \ v'(0) = 0, \tag{29}$$

where the exact solution of Eq. (28) is $v(x) = x^3$.

Let us consider $m = 3$, then we have

$$v(x) = \sum_{i=0}^{3} c_i T^*(x) \tag{30}$$

where the first term of the series is halved.

$$v(x) = \frac{1}{2} c_0 T_0^*(x) + c_1 T_1^*(x) + c_2 T_2^*(x) + c_3 T_3^*(x)$$

$$v(x) = \frac{1}{2}c_0(1) + c_1(2x-1) + c_2(8x^2-8x+1) + c_3(32x^3-48x^2+18x-1) \tag{31}$$

Now applying initial conditions (29)

$$v(0) = \frac{1}{2}c_0 - c_1 + c_2 - c_3 = 0 \tag{32}$$

Again applying initial conditions (29)

$$v'(0) = c_1 - 4c_2 + 9c_3 = 0 \tag{33}$$

Now from definition of Caputo fractional derivative (Theorem 1) and Eq. (23), we get

$$18.0544c_2x^{0.5} + c_3(144.4320x^{1.5} - 108.3264x^{0.5}) = 3x^3 + \frac{8}{\sqrt{\pi}}x^{\frac{3}{2}} \tag{34}$$

In our case, a particular choice for the collocation points is $x = 0.75, 0.25$ then Eq. (34) becomes

$$15.6356c_2 - 0.0021c_3 = 4.1972 \tag{35}$$

$$9.0272c_2 - 36.1092c_3 = 0.6111 \tag{36}$$

Now combining Eqs. (32), (33), (35) and (36), we get

$$c_0 = 0.6250, c_1 = 0.4688, c_2 = 0.1875, c_3 = 0.0313.$$

After putting the constants c_0, c_1, c_2, c_3 in Eq. (31), we get $v(x) = x^3$, which is same as the exact solution. Bhrawy and Alofi [26] solved the same problem using operational matrix for fractional integration based on shifted Chebyshev polynomials and got exact solution.

Example 2 The following fractional order IVP [27]

$$D^{\alpha}v(x) + v(x) = x^2 - x + \frac{2}{\Gamma(3-\alpha)}x^{2-\alpha}, \tag{37}$$

subject to the initial condition

$$v(0) = 0, \ v'(0) = -1, \tag{38}$$

whose exact solution is $v(x) = x^2 - x$.

The approximate solution is evaluated at $\alpha = 1.5$ and different choices of m. Numerical results of Example 2 reveal that the Chebyshev collocation method is achieved best accuracy and faster convergence and compared with Deng [27] and

Table 1 Absolute error between the exact and approximate solutions for different values of $m =$ for Example 2

x	Absolute error $m = 3$	Absolute error $m = 4$	Absolute error $m = 5$
0	0	0	0
0.1	2.22e−14	3.88e−19	0
0.2	8.18e−14	1.26e−18	0
0.3	1.67e−13	2.26e−18	0
0.4	2.69e−13	3.16e−18	0
0.5	3.75e−13	3.81e−18	0
0.6	4.76e−13	4.22e−18	0
0.7	5.59e−13	4.48e−18	0
0.8	6.15e−13	4.80e−18	0
0.9	6.32e−13	5.51e−18	0

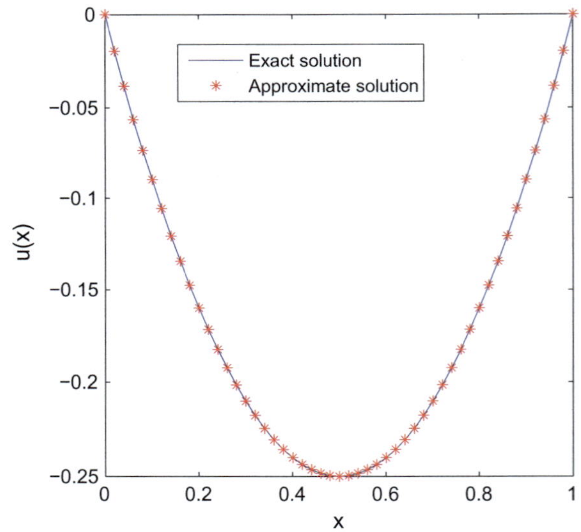

Fig. 1 Comparison with exact and approximate solution with $m = 5$ for Example 2

Bhrawy [28]. For satisfactory results, we need only small number of shifted Chebyshev basis functions; see in Table 1. In Fig. 1, the comparison between analytic solution and approximate solution are given. The absolute error is given in Table 1 for different values of m.

Example 3 Consider the fractional order IVP [29]

$$D^{(2.2)}v(x) + 1.3D^{(1.5)}v(x) + 2.6v(x) = \sin(2x), \tag{39}$$

with the initial conditions

Table 2 Comparison for different values of m with series solution [29] for Example 3

x	[29]	$m = 10$	$m = 15$	$m = 20$
0	0	0	0	0
0.1	0.0001478	0.0001543	0.0001493	0.0001483
0.2	0.0012750	0.0012920	0.0012790	0.0012770
0.3	0.0043990	0.0044310	0.0044110	0.0044070
0.4	0.0104100	0.0104800	0.0104500	0.0104400
0.5	0.0199600	0.0201400	0.0201000	0.0201000
0.6	0.0334500	0.0338900	0.0338500	0.0338400
0.7	0.0509200	0.0519500	0.0519000	0.0518900
0.8	0.0720400	0.0742300	0.0741800	0.0741700
0.9	0.0960400	0.1004000	0.1003000	0.0100300

Table 3 Absolute error for different values of m with series solution [29] for Example 3

x	Absolute error $m = 10$	Absolute error $m = 15$	Absolute error $m = 20$
0	0	0	0
0.1	6.5020e−6	1.5590e−6	5.2560e−7
0.2	1.7050e−5	4.0650e−6	1.7150e−6
0.3	3.2310e−5	1.1830e−5	8.2260e−6
0.4	7.0380e−5	4.2620e−5	3.7820e−5
0.5	0.0001754	0.0001412	0.0001353
0.6	0.0004387	0.0003988	0.0003920
0.7	0.0010230	0.0009780	0.0009704
0.8	0.0021930	0.0021440	0.0021350
0.9	0.0043510	0.0042990	0.0042900

$$v(0) = v'(0) = v''(0) = 0. \tag{40}$$

Equation (39) has series solution (see, [29]). The numerical solution obtained by proposed method is given in Tables 2 and 3 and Fig. 3. In Table 2, the comparison between series solution [29] and approximate solution and in Table 3, the absolute error are given for different choices of m. In Fig. 2, the comparison between series solution [29] and approximate solution is given for $m = 15$.

Example 4 Consider the equation

$$D^{\alpha} v(x) + v(x) = x^4 - \frac{1}{2}x^3 - \frac{3}{\Gamma(4-\alpha)}x^{3-\alpha} + \frac{24}{\Gamma(5-\alpha)}x^{4-\alpha} \tag{41}$$

subject to the initial conditions

$$v(0) = 0, \tag{42}$$

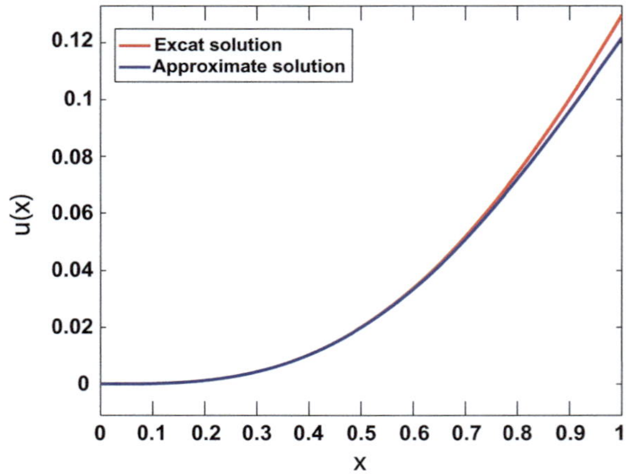

Fig. 2 Comparison between series solution [29] and our solution with $m = 15$ for Example 3

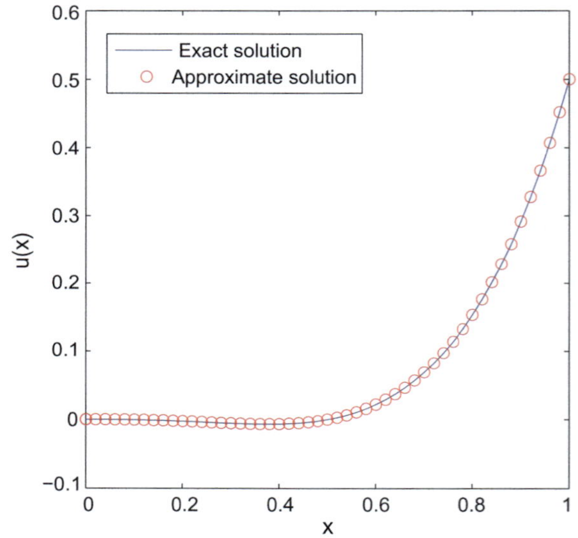

Fig. 3 Exact and approximate solutions with $m = 15$ for Example 4

where the exact solution is given by $v(x) = x^4 - \frac{1}{2}x^3$.

We presented absolute error for different values of m and $\alpha = 0.5$; see in Table 4. Also, we compared proposed method with Bhrawy [28] which indicates that the proposed method converges faster for acceptable solution and we require few terms of shifted Chebyshev polynomials.

Example 5 Consider the nonlinear fractional IVP [11]

$$D^3 v(x) + D^{2.5} v(x) + v^2(x) = x^4, \quad x \in (0, 1), \tag{43}$$

Table 4 Absolute error for different values of m for Example 4

x	Absolute error $m = 5$	Absolute error $m = 10$	Absolute error $m = 15$
0	0	0	0
0.1	5.05e−14	4.78e−14	4.88e−14
0.2	8.37e−14	9.78e−14	8.48e−14
0.3	1.05e−13	1.19e−13	1.12e−13
0.4	1.21e−13	1.26e−13	1.22e−13
0.5	1.34e−13	1.46e−13	1.39e−13
0.6	1.48e−13	1.68e−13	1.51e−13
0.7	1.65e−13	1.66e−13	1.65e−13
0.8	1.84e−13	1.52e−13	1.57e−13
0.9	2.06e−13	1.84e−13	1.70e−13

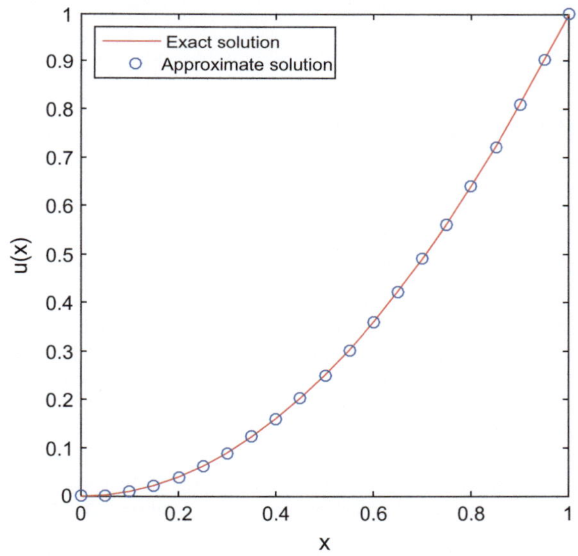

Fig. 4 Comparison between analytic and approximate solutions for Example 5

$$v(0) = v'(0) = 0, \ v''(0) = 2, \tag{44}$$

The exact solution is $v(x) = x^2$. We applied the proposed method which is described in Sect. 4, with $m = 3$, and we get the exact solution $v(x) = x^2$. This problem was also solved by Doha [30], and using Chebyshev spectral method both get the same result. The comparison between approximate solution and analytic solution is given in Fig. 4.

7 Conclusion

The numerical results obtained by the proposed method for solving multi-term FDEs are presented for different values of m. The shifted Chebyshev polynomials of the first kind are used to reduce multi-term FDE into system of equations, which is solved numerically. The Caputo fractional derivatives are considered in this problem. The error analysis of the derived formula and convergence analysis are discussed. For validation of proposed method, we compared numerical results with Bhrawy [28] in Examples 2, 4 which shows that the present scheme is more accurate. In Example 1, the proposed method and method [26] get the same result, and also in Example 5, the proposed method and method [30] get the same result. We obtained all computational results in MATLAB.

Acknowledgements The authors are very grateful to Department of Applied Mathematics & Humanities, S.V. National Institute of Technology, Surat, India, for providing Senior Research Fellowship.

References

1. Magin, R.L.: Fractional Calculus in Bioengineering. Begell House Publishers (2006)
2. Miller, K.S., Ross, B.: An Introduction to the Fractional Calculus and Fractional Differential Equations. Wiley, New York (1993)
3. Ortigueira, M.: Introduction to fraction linear systems. Part 2: discrete-time case. IEE Proc. Vis. Image Signal Proc. **147**, 71–78 (2000)
4. Deng, J., Ma, L.: Existence and uniqueness of solutions of initial value problems for nonlinear fractional differential equations. Appl. Math. Lett. **23**(6), 676–680 (2010)
5. Kilbas, A.A., Srivastava, H.M., Trujillo, J.J.: Theory and Applications of Fractional Differential Equations. Elsevier, San Diego (2006)
6. Abdulaziz, O., Hashim, I., Momani, S.: Application of homotopy-perturbation method to fractional IVPs. J. Comput. Appl. Math. **216**(2), 574–584 (2008)
7. Yang, S., Xiao, A., Su, H.: Convergence of the variational iteration method for solving multi-order fractional differential equations. Comput. Math. Appl. **60**(10), 2871–2879 (2010)
8. Ray, S.S., Bera, R.K.: Solution of an extraordinary differential equation by adomian decomposition method. J. Appl. Math. **2004**(4), 331–338 (2004)
9. El-Sayed, A.M., El-Kalla, I.L., Ziada, E.A.: Analytical and numerical solutions of multi-term nonlinear fractional orders differential equations. Appl. Numer. Math. **60**(8), 788–797 (2010)
10. Odibat, Z., Momani, S., Xu, H.: A reliable algorithm of homotopy analysis method for solving nonlinear fractional differential equations. Appl. Math. Model. **34**(3), 593–600 (2010)
11. Jafari, H., Das, S., Tajadodi, H.: Solving a multi-order fractional differential equation using homotopy analysis method. J. King Saud Univ. Sci. **23**(2), 151–155 (2011)
12. Diethelm, K., Ford, N.J., Freed, A.D.: A predictor-corrector approach for the numerical solution of fractional differential equations. Nonlinear Dynam. **29**(1), 3–22 (2002)
13. Kumer, P., Agrawal, O.P.: An approximate method for numerical solution of fractional differential equations. Signal Proc. **86**(10), 2602–2610 (2006)
14. Saadatmandi, A., Dehghan, M.: A new operational matrix for solving fractional-order differential equations. Comput. Math. Appl. **59**(3), 1326–1336 (2010)
15. Canuto, C., Hussaini, M.Y., Quarteroni, A., Zang, T.A.: Spectral Methods in Fluid Dynamics. Springer, New York (2012)

16. Doha, E.H.: Bhrawy AH (2008) Efficient spectral-Galerkin algorithms for direct solution of fourth-order differential equations using Jacobi polynomials. Appl. Numer. Math. **58**(8), 1224–1244 (1988)
17. Doha, E.H., Bhrawy, A.H.: Jacobi spectral-Galerkin method for the integrated forms of fourth-order elliptic differential equations. Numer. Methods Partial Differ. Equ. **25**(3), 712–739 (2009)
18. Doha, E.H., Bhrawy, A.H., Hafez, R.M.: A Jacobi-Jacobi dual-Petrov-Galerkin method for third- and fifth-order differential equations. Math. Comput. Model. **53**(9), 1820–1832 (2011)
19. Doha, E.H., Bhrawy, A.H., Ezz-Eldien, S.S.: Efficient chebyshev spectral methods for solving multi-term fractional orders differential equations. Appl. Math. Model. **35**(12), 5662–5672 (2011). http://dx.doi.org/10.1016/j.apm/2011.05.011
20. Doha, E.H., Bhrawy, A.H., Ezzeldeen, S.S.: Efficient Chebyshev spectral methods for solving multi-term fractional orders differential equations. Appl. Math. Model. **35**(12), 5662–5672 (2011)
21. Bhrawy, A.H., Alofi, A.S., Ezzeldeen, S.S.: A quadrature tau method for variable coefficients fractional differential equations. Appl. Math. Lett. **24**(12), 2146–2152 (2011)
22. Sweilam, N.H., Khader, M.M., Mahdy, A.M.S.: Numerical studies for fractional-order logistic differential equation with two different delays. J. Appl. Math. (2012)
23. Podlubny, I.: Fractional Differential Equations. Academic Press, New York (1999)
24. Diethelm, K.: The Analysis of Fractional Differential Equations. Springer, An Application-Oriented Exposition Using Differential Operators of Caputo Type (2010)
25. Mason, J.C., Handscomb, D.C.: Chebyshev Polynomials. Boca Raton. Chapman and Hall, New York, NY, CRC (2003)
26. Bhrawy, A.H., Alofi, A.S.: The operational matrix of fractional integration for shifted Chebyshev polynomials. Appl. Math. Lett. **26**(1), 25–31 (2013)
27. Deng, W.: Short memory principle and a predictor-corrector approach for fractional differential equations. J. Comput. Appl. Math. **206**(1), 174–188 (2007)
28. Bhrawy, A.H., Tharwat, M.M., Yildirim, A.: A new formula for fractional integrals of Chebyshev polynomials: application for solving multi-term fractional differential equations. Appl. Math. Model. **37**(6), 4245–4252 (2013)
29. Mdallal, Q.M., Syam, M.I., Anwar, M.N.: A collocation-shooting method for solving fractional boundary value problems. Commun. Nonlinear Sci. Numer. Simul. **15**(12), 3814–3822 (2010)
30. Doha, E.H., Bhrawy, A.H., Ezz-Eldien, S.S.: A Chebyshev spectral method based on operational matrix for initial and boundary value problems of fractional order. Comput. Math. Appl. **62**(5), 2364–2673 (2011)

Analysis of Typing Pattern in Identifying Soft Biometric Information and Its Impact in User Recognition

Soumen Roy, Utpal Roy and D. D. Sinha

Abstract As of now, the performance of keystroke dynamics biometric in user recognition is not acceptable in practice due to intra-class variations, high failure to enroll rate (FER) or various troubles in data acquisition methods or diverse use of sensing devices. As per the previous study, the performance of this technique can be improved by incorporation of gender information, a soft biometric characteristic, extracted from the typing pattern on a computer keyboard that provides some additional information about the user. This soft biometric trait has low user discriminating power but can be used to enhance the performance of user recognition in accuracy and time efficiency. Furthermore, it has been observed that the age group (18–30/30+ or <18/18+), gender (male/female), handedness (left-handed/right-handed), hand(s) used (one hand/both hands), typing skill (touch/others), and emotional states (anger/excitation) can be extracted from the way of typing on a computer keyboard for single predefined text. In this paper, we are interested in identifying multiple soft biometric traits using two leading machine learning methods: support vector machine with radial basis function (SVM-RBF) and fuzzy-rough nearest neighbor with vaguely quantified rough set (FRNN-VQRS) on multiple publicly available authentic and recognized keystroke dynamics datasets collected through a computer keyboard as well as touchscreen phone. The performance of machine learning methods are changed significantly in changing dataset in keystroke dynamics domain, but the evaluation performance of FRNN-VQRS in our experiment is promising and consistent in identifying traits. At the end, the impacts of the incorporation of soft biometric traits with primary biometric characteristics in user recognition are presented and compared the evaluation performance of nine anomaly detectors.

S. Roy (✉) · D. D. Sinha
University of Calcutta, Kolkata 700009, WB, India
e-mail: soumen.roy_2007@yahoo.co.in

D. D. Sinha
e-mail: devadatta.sinha@gmail.com

U. Roy
Visva-Bharati, Santiniketan 731235, WB, India
e-mail: roy.utpal@gmail.com

P. Chandra et al. (eds.), *Information Technology and Applied Mathematics*,
Advances in Intelligent Systems and Computing 699,
https://doi.org/10.1007/978-981-10-7590-2_5

Keywords Keystroke dynamics · Soft biometric · Machine learning
Fuzzy-rough NN (FRNN) · Vaguely quantified rough set (VQRS) · LIBSVM
Anomaly detector

1 Introduction

Cloud computing-based technologies (e-mail services, social network services, storage services, application services, Web hosting services, TV services, and more) are becoming very popular. As of now, only knowledge-based (PIN, password, graph pattern analysis) user authentication methods are common and used in cloud-based technology. But people are very lazy while choosing a healthy PIN, password, or graph pattern. As a result, this technique suffers from various attacks such as shoulder surfing attack, dictionary attack, and brute-force attack. Therefore, a higher level of the low cost, convenient, and efficient security solution is always demanded. Motor behavior information with the existing knowledge pattern in a particular way is the recent research field in access control system or information security. Keystroke dynamics technique is one of them.

Keystroke dynamics with the combination of knowledge-based authentication as a security solution could be used in practice. Since, keystroke dynamics is a method where the user can be identified by the way they type instead of what they type. Similar to handwriting recognition. It is better to choose daily used common words to get the habitual typing patterns which are more consistent [1], and it will decrease the false rejection rate (FRR). It has been established that habitual typing pattern is a behavioral biometric trait in biometric science that relates the issues in user identification/authentication. Nevertheless, being nonintrusive and cost-effective nature, keystroke dynamics is a strong alternative to other biometric modalities that can be easily integrated into any existing knowledge-based user authentication system with minor alternation.

Obtained accuracies in previous studies are impressive but not acceptable in practice due to the high failure to enroll rate (FER). The performances of behavioral biometrics are not impressive than morphological biometric modalities such as face print, fingerprint, and iris. It is not easy to achieve the acceptable accuracy. As per European standards, access control system mandates the false acceptance rate (FAR) is 1% and miss rate (FRR) is 0.001% [2]. As a behavioral biometric characteristics, keystroke dynamics is in trouble due to high rate of intra-class variations (problems in aging, mental state while typing, hand injury, tiredness, ...) or data acquisition methods (cross-device validation, timing resolution of the system, features selection, keyboard position, hand(s) used, ...) which increase the error rate in keystroke dynamics user authentication technique.

Recent keystroke dynamics studies [3, 4] found that personal traits such as age, gender, dominant hand, hand(s) used, and emotional state can be explained by the typing pattern on a computer keyboard. A study [5] found age and gender information can be explained through the behavior on the touchscreen phone. The science behind this technique is users' physical structure, hand weight, fingertips size, neurophysiological and neuropsychological factors reflect on the keyboard which discriminates the typing pattern.

These personal traits affect the typing characteristics and consequently affect the classification performance. For instance, touch size area of the child user might be less than the touch type of the adult user. Additionally, the length of the fingers of the female users might higher than fingers of male users. The right-handed user might type digraph consisting keys from the right side of the keyboard more quickly than type digraph consisting keys from the left side. Typing a digraph on a keyboard then taking a key from the left side and another from the right side of the keyboard, then typing pattern might differ among the users who used one hand or both hands. If the user is distracted or frustrated by the unnatural behavior of computer, then user's typing pattern changes massively. These are the soft biometric features which affect the typing characteristics consequently with the classification performance. Type of text, length, clock resolution of the system, the number of running software(s) and type, keyboard type and size with different layout also affect the performance. Therefore, the experimental results are impressive in the lab-based environment to be used in Web-based applications, but the performance of keystroke dynamics is not 100% accurate in practice. To improve the performance, inclusions of personal traits such as age, gender, handedness, hand(s) used, and typing skill are the new direction of keystroke dynamics research. Predicting of such traits is the new research direction not only to increase the keystroke dynamics user recognition performance; it has separate advantages in social network sites to e-business. The performance in identifying the personal traits is more important. Otherwise, classification performance might be decreased instead.

No study has been conducted to improve the performance of predicting personal traits to be used this technique in real-life applications with the Web-based environment as per our knowledge. The studies listed in Table 1 went step to identify the traits of the users based on typing pattern on a computer keyboard but not provided sufficient evidence to be used it on a touchscreen smartphone which is the most common and popular electronic gadgets. Social networking is becoming more popular to keep touch with the large and diverse body of people and groups. Nowadays, users of social network under age group below 18 are rapidly increasing. They easily reach out the contents which are not supposed to access or not suited for them. They share their personal information with strangers. Most of them have included a large number of strangers in their friend list. But no potential method has been applied to identify the age group and gender automatically based on the user's typing pattern on a computer keyboard and touchscreen smartphone instead of taking the age and gender information based on trust. The study [6] proposed that child user can be identified by typing style.

The journey of keystroke dynamics has been started in 1980 by Gains et al. [7]. Throughout these three decades, more than 500 papers have been published in the form of a journal, conference proceeding, and thesis, and still, the accuracy of this technique is not reached its goal. More research work has to be done so this technique can achieve its goal and can be used in practice. Ancillary information

Table 1 Success achieved by researchers to recognize the soft biometric traits on keystroke dynamics datasets

Study	Soft biometric traits	Categorization of classes	Obtained results	Instances	Test options	Classifiers	Simulators	Text patterns	Sensing devices
Epp et al. [4]	Emotional states	Anger and excitation	84% accuracy	1129	Tenfold cross-validation	C4.5	Weka [9]	Randomly selected piece of text from Alice's adventures in wonderland	Standard keyboard
Giot et al. [10]	Gender	Male and female	91% accuracy	7555	Fivefold cross-validation	SVM (RBF)	Python script	"greyc laboratory"	AZERTY keyboards
Syed-Idrus et al. [11]	Gender	Male and female	65–90% accuracy	2200	50% training ratio	SVM (RBF)	Python script	"leonardo dicaprio", "the rolling stones", "Michael schumacher", "red hot chilli peppers", "United States of America"	AZERTY and QWERTY keyboard
Syed-Idrus et al. [11]	Hand(s) used	One hand and two hands	90% accuracy		Do	Do	Do	Do	Do
Syed-Idrus et al. [11]	Age group	<30 and ≥ 30 <32 and ≥ 32	65–82% accuracy		Do	Do	Do	Do	Do
Syed-Idrus et al. [11]	Handedness	Left-handed and right-handed	70–90% accuracy		Do	Do	Do	Do	Do
Uzun et al. [6]	Age group	≤ 18 and 18+	$<10\%$ EER	500	Fivefold cross-validation	LDA, SVM (Linear, SVM (RBF)), etc.	MATLAB	".tie5Roanl", "Mercan Otu", ".tie5RoanlMercan Otu"	QWERTY keyboard

can significantly improve the recognition performance of biometric systems [8]. The details of the previous studies with inclusion of soft biometrics and research constraints they have used are summarized in Table 1. Studies conducted on different datasets to meet the aim of extracting the personal traits are based on typing pattern instead of improving the evaluation performed in identifying traits.

The keystroke dynamics datasets and the data acquisition methods are summarized in Table 2. All the data collection methods have been conducted in the lab-based environment. They used AZERTY and/or QWERTY keyboard as a sensing device. Text patterns of different studies are varied. Some of the texts are short, where some of the texts are long. Some of the texts are simply common words, where some of the texts are logically complex. Some of the studies used fivefold cross-validation test option, where some of the studies used training-testing ration test option in performance evaluation. The number of examples used in the previous studies is different, and some of them maintained the session in data acquisition.

Machine learning technique as a classification method is common in all the listed studies in Table 1. The selection of appropriate method is an important issue in keystroke dynamics domain, where the performance of one method jumps from 65 to 90%. In our study, we have applied FRNN with VQRS in identifying different traits and it has been observed that in all cases, FRNN with VQRS is proved suitable, very impressive, consistent, and significantly better than the previously used leading machine learning method SVM. In this study, we compare our approach with SVM.

The main objective of this study is twofold—first is to develop a model allowing to identify the proper gender, age group, handedness, and typing skill of users through the typing pattern on keyboard and identify the gender and age group through touchscreen phone for a predefined text, and the second one is to improve the accuracy by incorporating this soft biometric information as extra features in keystroke dynamics user recognition system with a suitable machine learning method.

Our objective and contribution of this paper are listed below:

- This study provides an efficient approach in identifying ancillary information through typing pattern on a computer keyboard as well as a touchscreen phone. The performance in our approach is promising than other approaches in the literature.
- Evaluate the performance of leading machine learning approaches to determine the multiple soft biometric information on multiple shared and authentic datasets.
- Evaluate and compare the performances of nine leading anomaly detectors in user recognition with and without incorporation of combinations of multiple soft biometric traits.

We have used authentic recognized and shared CMU keystroke dynamics dataset [12] along with dataset collected through an Android handheld device [13]. The details of the datasets we have used in our experiment are described in Table 3.

Table 2 Details of static and shared datasets on keystroke dynamics through keyboard

Study	Considered text	Number of subjects	Session	Repetition	Total sample	Feature subsets[a]	Sensing device
Uzun et al. [6]	".tie5Roanl", "Mercan Otu", ".tie5RoanlMercan Otu"	100	1	5	500, 500, 500	KD, DD, UD	QWERTY keyboard
El-Abed et al. [6]	"rhu.university"	51	3	15–20	951	DD, DU, UD, UU	Nokia Lumia 920
Antal et al. [14]	".tie5Roanl", "kicsikutyatarka", "Kktsf2!2014"	54	2	>30	3303, 3323, 3308	DU, DD, UD, pressure, velocity, total time, accelera-tions, total distance, etc.	Nexus 7 tablet, Mobil LG Optimus L7 II P710
Loy et al. [15]	"try4-mbs"	100	10	1	10000	Latency and pressure	A special pressure sensitive keyboard
Giot et al. [10]	"greyc laboratory"	100/133	12	4	4800	DD, UU, UD, DU	AZERTY keyboard
Roth et al. [16]	First paragraph of "A Tale of Two Cities" by Charles Dickens, Half page e-mail	45/50	2	4	90 audio files	Digraph time from keystroke sound	QWERTY keyboard
Killourhy et al. [12]	".tie5Roanl"	51	8	50	20400	KD, DD, UD	QWERTY keyboard
Jugurta et al. [17]	"chocolate", "zebra", "banana", "taxi", "computador calcula"	10, 8, 14, 15	2	5	100, 80, 140, 150	DD	Standard 101/102 keys, Brazilian lay-out—similar to the EUA layout
Bello et al. [18]	15 Spanish sentences and 15 Unix commands	58	1	1	282020 key events	Digraph	Uncontrolled environment
Syed-Idrus et al. [19]	"leonardo dicaprio", "the rolling stones", "Michael schumacher", "red hot chilli peppers", "United States of America"	110	2	10 for each hand(s) used classes	2200, 2200, 2200, 2200, 2200	UU, DU, UD, KD	AZERTY and QWERTY keyboard
Giot et al. [20]	Login information and password	118	3	1	9087 genuine samples, 10043 imposter samples, and 9346 imposed samples	UU, DD, UD, KD, total time	Uncontrolled

[a] KD, DD, UU, UD, and DU are described in the Eqs. 1–8

Table 3 Details of static and shared datasets used in this study

Dataset name	Study	Considered text	Gender	Age group	Handedness	Skill	Hand(s) used
Dataset A	Killourhy et al. [12]	".tie5Roanl"	M=38 F=27	<30=38 30+ = 27	L=9 R=56	T=37 O=28	–
Dataset B	Killourhy et al. [12]	"4121937162"	M= 25 F=13	<30=23 30+ = 15	L=4 R=38	T=20 O=18	–
Dataset C	Killourhy et al. [12]	"hester"	M=21 F=21	<30=24 30+ = 18	L=5 R=33	T=23 O=19	–
Dataset D	Idrus et al. [11]	"leonardo discaprio"	F=78 M=32	<30=51 ≥30=59	L=12 R=98	–	One hand=110 Both hands=110
Dataset E	Idrus et al. [11]	"the rolling stons"	F=78 M=32	<30=51 ≥30=59	L=12 R=98	–	One hand=110 Both hands=110
Dataset F	Idrus et al. [11]	"Michael schumaclur"	F=78 M=32	<30=51 ≥30=59	L=12 R=98	–	One hand=110 Both hands=110
Dataset G	Idrus et al. [11]	"red hot chilli perpers"	F=78 M=32	<30=51 ≥30=59	L=12 R=98	–	One hand=110 Both hands=110
Dataset H	Idrus et al. [11]	"United States of America"	F=78 M=32	<30=51 ≥30=59	L=12 R=98	–	One hand=110 Both hands=110
Dataset I	El-Abed et al. [13]	"rhu.university"	M=26 F=25	<18=11 19+ = 40	–	–	–
Dataset J	Uzun et al. [6]	".tie5Roanl"	–	≤18=51 18+ = 49	–	–	–
Dataset K	Uzun et al. [6]	"Mercan Otu"	–	≤18=51 18+ = 49	–	–	–
Dataset L	Uzun et al. [6]	".tie5RoanlMercan Otu"	–	≤18=51 18+ = 49	–	–	–

M, F, L, R, T, and O represent male, female, left-hander, right-hander, touch, and another type, respectively. Here, all the datasets were collected through a computer keyboard except Dataset I

2 Static and Shared Keystroke Dynamics Datasets

Many datasets on keystroke dynamics have been created in the last 30 years, but some of them are available on the Internet, or we can download on request. Details of the publicly available datasets are summarized in Table 2.

Soft biometric information is not included in all the datasets. Datasets created by Killourhy et al. [21], Idrus et al. [22], Yuzun et al. [6], and El-Abed et al. [13] provided soft biometric information with keystroke dynamics datasets which will be the most suitable datasets for our experiment. We have given some names on each dataset depending on considered text type and data acquisition method in this paper, so we can easily identify each dataset throughout this paper. Details are in Table 3.

3 Research Methodology

The proposed methodology has been described in the following subsections.

We have collected the publicly available authentic and recognize datasets which are most suitable for our experiment. The data acquisition protocol, number of session and repetition in data acquisition method, typing pattern, number of samples taken from a number of subjects, and the distribution of different categories of subjects are already summarized in Tables 2 and 3. We used the following equations to extract the features from the selected datasets, where some of the features were not presented in the dataset. We recalculated all the eight features by the following equations:

The timing features of the keystroke dynamics are as follows:

$$\text{Key} - \text{Duration (KD)} = R_i - P_i \tag{1}$$

$$\text{Up} - \text{Up Key Latency (UU)} = R_{i+1} - R_i \tag{2}$$

$$\text{Down} - \text{Down Key Latency (DD)} = P_{i+1} - P_i \tag{3}$$

$$\text{Up} - \text{Down Key Latency(UD)} = P_{i+1} - R_i \tag{4}$$

$$\text{Down} - \text{Up Key Latency (DU)} = R_{i+1} - P_i \tag{5}$$

$$\text{Total} - \text{Time (t)} = R_n - P_1 \tag{6}$$

$$\text{Tri} - \text{graph Latency (T)} = R_{i+2} - P_i \tag{7}$$

$$\text{Four} - \text{graph Latency (F)} = R_{i+3} - P_i \tag{8}$$

Here, P and R represent the pressed and released time of each ith key event while typing a pre-defined text. We have used a different combination of features to find the best choice of feature subset. Generally speaking, we have not applied any filtered or wrapper approach to select the features, and we have not considered any gyroscope, acceleration, acceleration with gravity, proximity, key pressure, and fingertips size information. We normalized all the datasets within the range [−1, +1] in order to speed up the process.

We have used two leading machine learning approaches: SVM and FRNN. Fuzzy-rough nearest neighbor (FRNN) [23] classification algorithm is an alternative to Sarkar's fuzzy-rough ownership function (FRNN-O) approach [24]. In SVM, we have used RBF with the cost function ($C = 128$, $\gamma = 0.125$) as per guided by [25]. In FRNN with VQRS, we have used the nearest neighbor, k=6, and the similarity function described below.

$$R_a(x, y) = \exp\left(-\frac{(a(x) - a(y))^2}{2\sigma_a^2}\right) \tag{9}$$

where σ_a^2 is the variance of feature a, and $R_a(x, y)$ is the degree of which objects x and y are similar for feature a.

To get the impact of soft biometric traits, each sample is manually labeled with the score 0 and 1 for the real different gender, age group, handedness, and typing skill of the user.

The next stage is to evaluate the nine anomaly detectors using the typing pattern and manually created soft biometric traits to get the impact and the importance of each trait and the combination of traits. The same procedures were used to train and test the samples. Same evaluation methodology was followed as per the study [12] described except median proximity instead of mean suggested to create the template of the user guided by the study [26]. The equal error rate (EER) was recorded for each user. The EER is a parameter to evaluate the performance of all anomaly detectors. EER is the value where FAR and FRR both are same.

4 Experimental Results

Two leading machine learning algorithms have been applied to each dataset, and accuracy with tenfold cross-validation has been listed in Tables 4, 5, 6, 7, 8, and 9 to predict the soft biometric information. As per obtained results, FRNN with VQRS is proved to be the suitable learning methods in both desktop and Android environments. Accuracies were recorded by Weka 3.7.4 simulator with the parameter values mentioned in the previous section.

The results recorded in Table 4 indicate that the gender can be predicted with the accuracy 83–95% by using FRNN-VQRS in different environments. The accuracies are compared with the previously used SVM by the study [11]. Recognising gender is an important soft biometric trait which is not only used to improve the accuracy in user recognition system but also could help to make a fake free user account or auto profiling the user.

Identifying the age group is an important issue in keystroke dynamics domain. Table 5 represents the accuracy in identifying the age group (<30/30+) on different datasets. The 75–95% accuracy was recorded in identifying the age group by our approach. Identifying the age group (<18/18) also was recorded to identify the child users based on their typing rhythm. So, we can protect the teens from the looming

Table 4 Accuracy with standard deviation in identifying gender (male/female)

Classification algorithms	Accuracy (%)								
	Dataset A	Dataset B	Dataset C	Dataset D	Dataset E	Dataset F	Dataset G	Dataset H	Dataset I
LibSVM	82.67 (0.21)	76.30 (0.69)	86.87 (0.46)	72.12 (0.79)	74.60 (0.43)	71.28 (0.46)	76.07 (1.25)	76.14 (0.66)	75.52 (1.10)
FRNN-VQRS	94.97 (0.12)	87.29 (0.48)	93.97 (0.31)	80.92 (1.03)	86.54 (0.79)	83.29 (0.86)	86.44 (1.13)	90.17 (1.23)	83.87 (1.42)

Table 5 Accuracy with standard deviation in identifying age group (<30/30+)

Classification algorithms	Accuracy (%)							
	Dataset A	Dataset B	Dataset C	Dataset D	Dataset E	Dataset F	Dataset G	Dataset H
LibSVM	81.97 (0.32)	73.57 (0.42)	83.49 (0.46)	62.03 (1.43)	63.58 (1.45)	60.08 (1.74)	65.87 (1.19)	64.82 (1.12)
FRNN-VQRS	94.93 (0.17)	87.23 (0.46)	93.90 (0.41)	75.97 (0.91)	81.95 (1.09)	79.05 (0.84)	84.90 (1.01)	88.61 (0.97)

Table 6 Accuracy with standard deviation in identifying age group (<18/18+)

Classification algorithms	Accuracy (%)			
	Dataset I	Dataset J	Dataset K	Dataset L
LibSVM	82.00 (1.56)	88.60 (1.77)	88.11 (1.97)	90.4 (1.98)
FRNN-VQRS	87.99 (1.00)	88.32 (4.13)	93.22 (3.43)	94.68 (3.00)

Table 7 Accuracy with standard deviation in identifying handedness

Classification algorithms	Accuracy (%)							
	Dataset A	Dataset B	Dataset C	Dataset D	Dataset E	Dataset F	Dataset G	Dataset H
LibSVM	91.61 (0.16)	92.46 (0.22)	94.85 (0.27)	96.93 (0.59)	97.05 (0.29)	95.97 (0.37)	95.65 (0.79)	96.55 (0.72)
FRNN-VQRS	97.77 (0.07)	94.99 (0.20)	97.12 (0.19)	98.90 (0.26)	98.85 (0.18)	98.11 (0.43)	98.11 (0.37)	98.44 (0.57)

Table 8 Accuracy with standard deviation in identifying typing skill (touch/others)

Classification algorithms	Accuracy (%)		
	Dataset A	Dataset B	Dataset C
LibSVM	83.72 (0.30)	76.91 (0.54)	87.03 (0.30)
FRNN-VQRS	95.19 (0.16)	88.63 (0.27)	93.94 (0.17)

Table 9 Accuracy with standard deviation in identifying hand(s) used

Classification algorithms	Accuracy (%)				
	Dataset D	Dataset E	Dataset F	Dataset G	Dataset H
LibSVM	96.93 (0.59)	97.05 (0.29)	95.97 (0.37)	95.65 (0.79)	96.55 (0.72)
FRNN-VQRS	98.90 (0.26)	98.85 (0.18)	98.11 (0.43)	98.11 (0.37)	98.44 (0.57)

Table 10 Comparative analysis of anomaly detectors with inclusion of personal traits by using performance metric equal error rate (EER) in % on Dataset A

Anomaly detectors	Including all features	Including age, gender, and handedness	Including age and gender	Including age group	Including gender	Including handedness	Including typing skill	Only typing pattern
Canberra	12.28	13.33	13.85	15.19	15.24	16.16	15.18	16.81
Chebyshev	2.8	4.41	5.27	7.94	7.9	9.65	7.9	11.3
Czekanowski	2.8	4.39	5.26	7.87	8.05	9.61	7.99	12.08
Lorentzian	**2.53**	**4.07**	**4.82**	**7.39**	**7.38**	**9.16**	**7.48**	**10.69**
Minkowski	4.44	6.84	8.18	12.02	11.98	14.42	11.74	17.18
Sorensen	2.8	4.39	5.26	7.84	8.01	9.6	7.96	11.92
Euclidean	3.85	6	7.15	10.55	10.39	12.46	10.26	14.71
Manhattan	2.8	4.41	5.27	7.94	7.9	9.65	7.9	11.3
SVM	3.59	5.2	6.24	9.04	9.12	11.1	9.37	10.64

threads on the Internet, or we can filter the content which is more appropriate for that category of users by implementing the autosensing firewall appropriate for the users.

Handedness of the user can be predicted as per the results recorded in Table 7. The 94–99% accuracy can be achieved in identifying the handedness of the users by their typing style.

Table 8 represents the results recorded in identifying the typing skill on variously shared datasets. The 89–95% accuracy was recorded. It can be used as soft biometric traits to improve the accuracy in user recognition through keystroke dynamics characteristics.

A number of hands used by the user can be predicted by the typing pattern, and the accuracy of identifying the hand(s) used is 99% as per the recorded results in Table 9. This trait was not considered as a soft biometric trait in the study [27].

The results recorded in identifying the soft biometric traits are impressive and significantly better than the previously used SVM in all cases.

Tables 10, 11, and 12 represent the performance of nine anomaly detectors described in [28] after incorporation of soft biometric information. We observed that instead of using the only gender, multiple soft biometrics information decreases the EER significantly. Here, we take the median of samples to create the template. In keystroke dynamics, domain median proximity is better than mean [26].

Table 10 represents the evaluation performance, EERs in % on Dataset A. The EER jumps from 10.69 to 2.53% by incorporating all the soft biometric traits mentioned for Dataset A. The improvement of EER does not depend on the text pattern as per the results recorded for the different texts in Tables 11 and 12. The performance of Lorentzian is impressive than the others anomaly detectors.

Table 11 Comparative analysis of anomaly detectors with inclusion of personal traits by using performance metric equal error rate (EER) in % on Dataset B

Anomaly detectors	Including all features	Including age, gender, and handedness	Including age and gender	Including age group	Including gender	Including handedness	Including typing skill	Only typing pattern
Canberra	4.43	5.74	6.23	7.62	8.13	9.36	7.75	10.03
Chebyshev	2.87	4.32	4.91	7.58	7.82	10.16	7.38	11.56
Czekanowski	2.87	4.34	4.86	7.65	8.05	10.2	7.8	12.48
Lorentzian	**2.73**	**4.1**	**4.6**	**7.33**	**7.47**	**9.87**	**7.06**	**11.2**
Minkowski	3.73	5.69	6.53	9.77	9.65	12.8	9.4	14.36
Sorensen	2.87	4.34	4.86	7.65	8.05	10.2	7.8	12.48
Euclidean	3.31	5.21	5.93	9.01	9.02	11.81	8.59	13.3
Manhattan	2.87	4.32	4.91	7.58	7.82	10.16	7.38	11.56
SVM	3.35	5.07	5.76	8.95	9.22	11.27	8.74	11.88

Table 12 Comparative analysis of anomaly detectors with inclusion of personal traits by using performance metric equal error rate (EER) in % on Dataset C

Anomaly detectors	Including all features	Including age, gender, and handedness	Including age and gender	Including age group	Including gender	Including handedness	Including typing skill	Only typing pattern
Canberra	22.08	23.73	24.29	26.27	26.54	27.67	26.19	28.16
Chebyshev	4.59	6.77	7.68	10.06	10.35	11.43	10.15	12.79
Czekanowski	4.56	6.75	7.43	9.81	10.06	11.36	10.05	13.24
Lorentzian	**4.41**	**6.6**	**7.53**	**9.76**	**10.21**	**11.34**	**10**	**12.69**
Minkowski	6.09	9.32	10.51	14.45	13.97	16.33	13.49	17.89
Sorensen	4.54	6.75	7.28	9.58	9.72	11.31	9.95	12.3
Euclidean	5.51	8.21	9.29	12.27	12.14	13.79	11.66	15.16
Manhattan	4.59	6.77	7.68	10.06	10.35	11.43	10.15	12.79
SVM	4.55	6.45	7.12	9.09	9.56	10.44	8.95	11.9

5 Conclusion

It is possible to predict the gender, age group, handedness, hand(s) used, and typing skill of the user through the way of typing as it is evident from our experiment with impressive results. Keystroke dynamics can be used to recognize the gender and age group prediction since keystroke dynamics is a common measurable activity to be used in Web-based applications. The activities on the keyboard and touchscreen are behavioral biometric characteristics, and it could be used to predict the gender and age group to deal with the problem of fake accounts and would enable to create a more loyal and authentic social networking sites. This may facilitate social network sites a fake free, genuine, and more loyal user base. Automatically identifying soft biometric traits and the incorporation of such traits also can be used as additional information to reduce the error rate in the keystroke dynamics user recognition system. This technique also could help e-commerce site to reach out to the right client. Similarly, this could avoid adverse products more efficiently based on the gender and age group. This technique also can be very useful in a Web-based environment for auto profiling of the users. The results also show that age group below 18 can be identified based on typing pattern which can be used to protect the teens from Internet threats.

We have used two leading machine learning methods to predict personal traits on multiple publicly available authentic datasets. Our proposed approach FRNN-VQRS, a new approach to FRNN, achieved impressive results significantly better than previously used SVM with RBF to determine the personality traits in desktop and Android environments. The evaluation performance of FRNN-VQRS is consistent. This is a very positive outcome in keystroke dynamics system for a single predefined text which can be used as soft biometric additional features in user identification/authentication technique which decrease the EER 10.69–2.53 on CMU keystroke dynamics dataset. This is the modest as well as an efficient approach toward the keystroke dynamics user authentication system.

The impact of incorporation of soft biometric traits with primary biometric characteristic has been presented and discussed. We evaluated nine anomaly detectors methods on CMU Keystroke dynamics datasets, and obtained results have been compared. This is the modest as well as an efficient approach to improve the performance of the keystroke dynamics user recognition system.

References

1. Roy, S., Roy, U., Sinha, D.D.: Free-text user authentication technique through keystroke dynamics. In: 2014 International Conference of High Performing Computer Application ICHPCA 2014 (2015)
2. CENELEC, alarm systems—Access control systems for use in security applications—part 1. In: System Requirements, EN 50133-1 edition (1996)
3. Zulkarnain, S., et al.: Soft biometrics for keystroke dynamics: profiling individuals while typing passwords. Comput. Secur. (2014)
4. Epp, C., Lippold, M., Mandryk, R.L.: Identifying emotional states using keystroke dynamics. In: Proceedings of SIGCHI Conference Human Factors Computer System, pp. 715–724 (2011)
5. Roy, S., Roy, U., Sinha, D.D.: ACO-random forest approach to protect the kids from internet threats through keystroke. Int. J. Eng. Technol. 2–9 (2017) (Accepted)
6. Uzun, Y., Bicakci, K., Uzunay, Y.: Could We Distinguish Child Users from Adults Using Keystroke Dynamics? (2014)
7. Gaines, R.S., Lisowski, W., Press, S.J., Shapiro, N.: Authentication by Keystroke Timing: Some Preliminary Results, in Technical Report R-2526-NSF. Rand Corporation, May (1980)
8. Jain, A., Nandakumar, K., Lu, X., Park, U.: Integrating faces, fingerprints, and soft biometric traits for user recognition. Biometr. Authenticat. no. May, 259–269 (2004)
9. Frank, E., Hall, M.A., Witten, I.H.: The Weka Workbench Data Mining Practical Machine Learning Tools and Techniques, 4th ed. (1999)

10. Giot, R., Rosenberger, C.: A new soft biometric approach for keystroke dynamics based on gender recognition. Int. J. Inf. Technol. Manag. Spec. Issue Adv. Trends Biometric. **11**(August), 1–16 (2012)
11. Idrus, S.Z.S., Cherrier, E., Rosenberger, C., Bours, P.: Soft biometrics for keystroke dynamics. In: 10th International Conference on Image Analysis and Recognition (ICIAR), pp. 11–18 (2013)
12. Killourhy, K.S., Maxion, R.A.: Comparing anomaly-detection algorithms for keystroke dynamics. In: Proceedings of the International Conference on Dependable Systems and Networks, pp. 125–134 (2009)
13. El-Abed, M., Dafer, M., El Khayat, R.: RHU keystroke: a mobile-based benchmark for keystroke dynamics systems. In: 2014 International Carnahan Conference Secured Technology, pp. 1–4 (2014)
14. Antal, M., Szabo, L.Z.: An evaluation of one-class and two-class classification algorithms for keystroke dynamics authentication on mobile devices. In: Proceedings of 2015 20th International Conference Control System Computer Science CSCS 2015, pp. 343–350 (2015)
15. Loy, C.C., Lim, C.P., Lai, W.K.: Pressure-based typing biometrics user authentication using the fuzzy ARTMAP neural network. In: International Conference on Neural Information Processing (ICONIP) (2005)
16. Roth, J., Liu, X., Ross, A., Metaxas, D.: Biometric authentication via keystroke sound. In: 2013 International Conference on Biometrics, pp. 1–8 (2013)
17. Montalvão, J., Freire, E.O., Bezerra, M.A., Garcia, R.: Contributions to empirical analysis of keystroke dynamics in passwords. Pattern Recogn. Lett. (2015)
18. Bello, L., Bertacchini, M.: Collection and publication of a fixed text keystroke dynamics dataset. In: CACIC 2010—XVI Argentino Ciencias LA Congress of Computer Science, pp. 822–831 (2010)
19. Idrus, S.Z.S., Cherrier, E., Rosenberger, C., Bours, P.: Soft biometrics database: a benchmark for keystroke dynamics biometric systems. In: 2013 International Conference of the Biometrics Special Interest Group (BIOSIG), 2013, no. September, pp. 1–8
20. Giot, R., El-Abed, M., Rosenberger, C.: Web-based benchmark for keystroke dynamics biometric systems: a statistical analysis. In: Intelligence of Information Hiding and Multimedia Signal Process, pp. 11–15 (2012)
21. Killourhy, K.S.: A scientific understanding of keystroke dynamics. Dr. Thesis, School of Computer Science, Computer Science Department Carnegie Mellon University Pittsburgh, PA 15213, no. January, pp. 1–198 (2012)
22. Idrus, S.Z.S., Cherrier, E., Rosenberger, C., Bours, P.: Soft biometrics database: a benchmark for keystroke dynamics biometric systems. In: 2013 International Conference of the Biometrics Special Interest Group (BIOSIG), September, pp. 1–8 (2013)
23. Jensen, R., Cornelis, C.: Fuzzy rough nearest neighbour classification and prediction. Theor. Comput. Sci. **412**(42), 5871–5884 (2011)
24. Sarkar, M.: Fuzzy-rough nearest neighbor algorithms in classification. Fuzzy Sets Syst. **158**(19), 2134–2152 (2007)
25. Hsu, C.-W., Chang, C.-C., Lin, C.-J.: A practical guide to support vector classification. BJU Int. **101**(1), 1396–1400 (2008)
26. Alghamdi, S.J., Elrefaei, L.A.: Dynamic user verification using touch keystroke based on medians vector proximity. In: Proceedings of 7th International Conference Computer Intelligence Communication System Networks, CICSyN 2015, pp. 121–126 (2015)
27. Zulkarnain, S., et al.: Keystroke Dynamics Performance Enhancement With Soft Biometrics
28. Roy, S., Roy, U., Sinha, D.D.: Comparative study of various features-mining-based classifiers in different keystroke dynamics datasets. In: Proceedings of First International Conference on Information and Communication Technology for Intelligent Systems, vol. 2, pp. 155–164 (2016)

An Ideal and Perfect (*t*, *n*) Multi-secret Sharing Scheme Based on Finite Geometry

Barun Duari and Debasis Giri

Abstract Secret sharing has numerous applications in cryptography field and distributed computing. Threshold secret allotment scheme has been well considered a lot for the last three decades and proposed many such efficient schemes. The main objective of allotment sharing scheme is to deliver the secret to some parties so that only desirable subsets of the parties can get back the secrets whereas secret cannot be leaked by other parties. Nowadays, an ideal scheme shares multiple secrets with perfect security is of high demand. In this paper, we have designed a multi-secret sharing scheme on the basis of geometry in Galois field. The scheme is ideal, perfect without information leakage.

Keywords Secret sharing · Multi-secret sharing · Information security Cryptography · Distributed computing

1 Introduction

In reality, we believe that only one authority may not trustworthy to secure the secret information. To overcome this, secret sharing technique is considered. In (t, n)-Threshold secret allotment scheme, a dealer is responsible for distributing n pieces of information called shares or shadows related with secret information to the n participants where secret to be protected among them in such a way that no group of participants having cardinality less than the predetermined threshold value t number of participants can cooperate to reform the secret and only when t or more number of participants can reconstruct the secret information.

B. Duari (✉)
Tamralipta Mahavidyalaya, Tamluk 721636, WB, India
e-mail: barunduari@gmail.com

D. Giri
Haldia Institute of Technology, Haldia 721657, WB, India
e-mail: debasis_giri@hotmail.com

P. Chandra et al. (eds.), *Information Technology and Applied Mathematics*,
Advances in Intelligent Systems and Computing 699,
https://doi.org/10.1007/978-981-10-7590-2_6

Threshold secret sharing scheme was first proposed by Shamir [1] and Blakley [2] independently in 1979. Shamir implemented the concept of Lagrange polynomial interpolation in his scheme, whereas the scheme by Blakley is constructed on the geometric approach that uses the notion of intersecting hyperplanes. Those schemes are very helpful in the management of cryptographic key. But the computation in Shamir's scheme is rigorous. Hence, it is not efficient to implement in practical. Another drawback is that the large memory space is needed to hold all the shares. Ramp secret sharing scheme has been proposed by Blakley and Meadows [3] in 1985 to diminish the length of shares than the scheme of Shamir. In the scheme of ramp, they considered intermediate sets from participants set, which are neither qualified nor banned sets to get back the secrets, and hence somewhat leak information about the secret. Many researchers have been proposed variety of secret sharing schemes such as single and multi-secret sharing schemes. But constructing an efficient provable secured, perfect and ideal scheme without leaking any information about secret or shares is a great challenge. The traditional (t, n) threshold secret allotment schemes [4–10] are particular case of general secret sharing scheme. Their qualified subsets are those that have at least a certain, predetermined number of participants. In 2006, Bai [11] designed a threshold secret allotment scheme using the notion of the projection of matrix. The scheme is designed on the basis of the invariance property of the projection of matrix. Tang and Yao [12] invented multi-prover with zero-knowledge arguments-based threshold scheme in 2008, and this scheme improves the security on multiparty computation protocol to prevent the malfeasances by unqualified set of participants. It checks for least possible required qualified participants before going to share the secret. In 2009, Shi and Zhong [13] proposed that the modification of the Shamir's threshold value is possible beyond dealer's assistance. In their scheme, all participants collaborate to acquire the act of dealer. Chou et al. [14] proposed a threshold scheme using Sudoku in 2010 with enhanced security in data transmission as a result of huge solutions in Sudoku. In 2010, Harn and Lin [15] extended the basic idea of a (t, n) secret sharing scheme and recommended a strong (n, t, n) verifiable secret sharing (VSS) scheme. In their scheme, each participant also can act as a dealer. In 2012, Liu et al. [16] proposed an efficient strong (n, t, n) VSS that was more powerful over Harn and Lin's scheme [15]. In addition, they proposed an (n, t, n) multi-secret sharing (MSS) scheme for sharing $n - t + 1$ secrets to participants. In 2013, Patil and Desmukh [17] proposed a scheme using the concept of dot product of linearly independent vectors. Guo et al. [18] designed a key-lock-pair mechanism-based secret sharing scheme with general access structures in the year of 2013. Tassa [19] recommended a novel, hierarchical threshold secret sharing scheme in 2007 using the Birkhoff interpolation where the secret is shared among a group of participants which is partitioned into some levels, and the respective access structure is determined by a sequence of threshold needs. In 2012, Zhao et al. [20] proposed a secret sharing scheme based on the properties of the Jordan matrix that can achieve the (t, n) threshold and the adversary access structure. That is, there are some subsets that contain at least t participants that cannot reconstruct the secret. Su et al. [21] proposed a multi-secret sharing scheme based on a hyper-spherical function in the year of 2007. Hsu et al. [22] proposed an ideal multi-secret sharing scheme based on MSP in 2011 where all n participants should be present to reconstruct the secret information. Subba and Chakravarthy [23] invent a novel secret sharing system that is designed on the basis of Chinese remainder theorem in the year of 2014. In 2016, Yuan et al. [24] designed a polynomial interpolation-based secret sharing scheme where threshold value is changeable.

Most of the above schemes we have seen that partially or fully leak information on shares or shadows. Some of the above schemes are not ideal or not perfect or high computation cost and also there used either polynomial interpolation or solving the system of linear equation to reform the secret directly or indirectly. In our proposed scheme, dealer considers a hyperplane in t-dimensional space and chooses a secret point on plane. Dealer also chooses a straight line that passes through the secret point and publishes this publicly. Now he/she selects n non-collinear points randomly and distributes the shadows of those to participants. When t participants will come together, they can construct the hyperplane and then reconstruct the secret intersection point of hyperplane using publicly known straight line.

The outline of the paper is structured as follows. We use some notations discussed in Sect. 2. In Sect. 3, we state some basic descriptions. We introduce our developed scheme at Sect. 4. Security measure of our proposed scheme is described in Sect. 5. Our scheme is compared with some of the above schemes in respect of such defined parameters is discussed in tabular form in Sect. 6 and we conclude in Sect. 7.

2 Notation

Here, we discuss some notation used in our scheme. $GF(p)$ defines a Galois field for large prime p and G_p is a cyclic group with generator g. The own secret/identity of dealer is denoted by k. Individual participant is denoted by P_j, and his/her corresponding own secret/identity is denoted by p_j where $j = 1, 2, \ldots, n$. Here, $x_i^{(s)}, x_i^{(p)}, x_i^{(q)}$ and $x_i^{(j)}$ represent the ith coordinates of points S, P, Q and share the jth participant P_j, respectively, in t-dimensional space. The shadow of share s_j is denoted by $\hat{s}_j$, $j = 1, 2, \ldots, n$.

3 Basic Description

Before going to discuss our proposed scheme, we discuss some basic concept in this section.

Threshold Secret Allotment Scheme: A (t, n) threshold secret allotment scheme is a mechanism of distribution of a secret among a set of n parties (denoted by P) where t $(\leq n)$ is the number of threshold with the following two requirements:

(1) No group of $t - 1$ participants can solve secret, even if they collaborate.

(2) When it requires that the secret information be reformed, a large enough number of parties ($\geq t$) can always do it.

The threshold value t is selected by the particular participant called dealer D. If D needs to share the secret key or information among the n participants those are in P, all participants share their respective secret. The share of each participant should be dispersed secretly. So, no participant knows about the share of any other participant.

Information Rate: The ratio of the *size* of the secret over the *size* of the largest share represents the information rate of a secret sharing.

Ideal Scheme: A scheme is ideal if the *information rate* is one; i.e., share of each participant has the same length of the secret.

Perfect: A $(t,\ n)$ threshold secret sharing scheme is perfect if any $(t-1)$ or lesser shares cannot reconstruct any information or fact about secret S.

4 Proposed Scheme

We choose a large prime number p and consider the Galois field $GF(p)$. Let us consider a t variants linear equation in that field that is a hyperplane in t-dimensional space and also consider a point S treated as secret on that hyperplane. Now, choose two points P, Q which are collinear with secret point S but not coplanar with the hyperplane and publish these points publicly. Now, dealer chooses n linear independent points on plane as shares for n participants. When any t participants will come together, they can reform the hyperplane and then get the intersection point of this linear equation and equation of the line PQ which is unique with secret point S. In our scheme, original share of each participant is also secured from rest of the world in use of shadow point of respective original share. So, no information is leaked about secret **S** and shares also. Our scheme consists of the following phases.

Initialisation Phase: In our scheme, dealer D chooses G_p and g (generator on G_p) where p is large prime number. D chooses any integer k form G_p and computes $\alpha = g^k$. D publishes public values $\{G_p, g, \alpha\}$ to a public directory. Each participant P_j chooses his/her own secret p_j and publishes $\beta_j\ (= g^{p_j})$ with his/her own identity to the dealer D.

Construction Phase:

(1) Dealer D chooses a linear equation in $GF(p)$,

$$\sum_{i=1}^{t} a_i x_i = c\ (mod\ p). \tag{1}$$

where $\forall a_i$ and c are nonzero.

The dealer chooses a secret point S $\left(x_1^{(s)}, x_2^{(s)}, \ldots, x_t^{(s)}\right)$ that satisfies (1). The secret S will be shared with n participants.

(2) D chooses two points P $(x_1^{(p)}, x_2^{(p)}, \ldots, x_t^{(p)})$ and Q $(x_1^{(q)}, x_2^{(q)}, \ldots, x_t^{(q)})$ those are collinear with S but not satisfies (1). P, Q are published publicly.

(3) D chooses a point s_j $(x_1^{(j)}, x_2^{(j)}, \ldots, x_t^{(j)})$ except S that satisfies (1) as actual share for each participant P_j and supplies shadow point $\hat{s}_j$ of s_j to the P_j for $j = 1, 2, \ldots, n$, where all those points are linearly independent.

- Compute $\hat{s}_j = s_j \cdot \beta_j^k$; for $j = 1, 2, \ldots, n$;

  where $s_j \cdot \beta_j^k$ is nothing but scalar (β_j^k) multiplication to the vector $(\vec{s}_j)$.
- Distribute $\hat{s}_j$ to corresponding participant.

Formation of s_j:

Rewrite Eq. (1) in parametric form

$$\begin{cases} x_1^{(j)} = \left(c - \sum_{i=2}^{t} r_{i-1}^{(j)}\right) / a_1 \\ x_i^{(j)} = r_{i-1}^{(j)} \end{cases} \quad \text{for } j = 1, 2, \ldots, n \text{ and } i = 2, 3, \ldots t$$

All r's are chosen by dealer randomly.

Recovery Phase:

(1) Each participant P_j recovers his/her own actual share point $s_j(x_1^{(j)}, x_2^{(j)}, \ldots, x_t^{(j)})$ from his/her shadow $\hat{s}_j$.

- Computes $s_j = \alpha^{-p_j}.\, \hat{s}_j$; for $j = 1, 2, \ldots, n$;

And assume

$$S = P + U(Q - P)$$

$$i.e.\ x_i^{(s)} = x_i^{(p)} + U\left(x_i^{(q)} - x_i^{(p)}\right) \text{ for } i = 1, 2, 3, \ldots t. \tag{2}$$

(2) When t or more participants will come together, they can reveal the equation as:

$$\begin{vmatrix} x_1 - x_1^{(1)} x_2 - x_2^{(1)} \ldots x_t - x_t^{(1)} \\ x_1^{(2)} - x_1^{(1)} x_2^{(2)} - x_2^{(1)} \ldots x_t^{(2)} - x_t^{(1)} \\ \ldots \\ \ldots \\ x_1^{(t)} - x_1^{(1)} x_2^{(t)} - x_2^{(1)} \ldots x_t^{(t)} - x_t^{(1)} \end{vmatrix} = 0$$

$$i.e \sum_{i=1}^{t} b_i x_i = d(mod\ p). \tag{3}$$

This equation is either unique or equivalent to (1). So, they are identical in graph.

(3) Now, compute using (2) and (3),

$$U = \left(d - \sum_{i=1}^{t} b_i x_i^{(p)} / \sum_{i=1}^{t} b_i (x_i^{(q)} - x_i^{(p)}) \right) \ (mod\ p). \tag{4}$$

So, we can reform the secret S $(x_1^{(s)}, x_2^{(s)}, \ldots, x_t^{(s)})$ from (2) and (4).

4.1 A Simple Example

Here, we illustrate a simple example of our proposed scheme. We consider (3, 4) threshold scheme in which 3 out of 4 participants can reconstruct the secret and assume $GF(11)$. Let dealer assume $k = 6$ as his/her own secret and publish $g = 2$, $\alpha = 9\ (mod\ 11)$. Now, the participants P_j $(j = 1, 2, 3, 4)$ choose their own secret say 3, 2, 1, 5, respectively, and send their corresponding β_j: 8, 4, 2, 10 to the dealer. Now, dealer assumes a linear equation:

$$3x_1 + 2x_2 + 5x_3 = 7\ (mod\ 11). \tag{1.1}$$

and also chooses a secret point S(2, 1, 2) satisfies the above plane. Again dealer chooses two points P(0, 5, 10) and Q(1, 3, 6) collinear with S but not satisfies the plane and publishes these two points publicly.

Now, dealer chooses four linearly independent points $s_1(3, 5, 2)$, $s_2(2, 2, 6)$, $s_3(1, 2, 0)$, $s_4(0, 1, 1)$ as actual share of participants, respectively, and distributes their shadow point use of corresponding β_j to P_j. Suppose for P_1, $\hat{s}_1 = s_1 \cdots \beta_1^k =(3, 5, 2)\ 8^6 =(9, 4, 6)$ (mod 11). Similarly, $\hat{s}_2 =(8, 8, 2)$; $\hat{s}_3 =(9, 7, 0)$; and $\hat{s}_4 =(0, 1, 1)$ for P_2, P_3, P_4, respectively. Now, distribute those shadows to corresponding participant.

Before reconstructing the secret, each P_j recovers his/her own actual share point s_j from his/her shadow $\hat{s}_j$ use of α and his/her secret p_j.

Suppose for $P_1, s_1 = \alpha^{-p_1}.\hat{s}_1 = 9^{-3}.\ (9,\ 4,\ 6) = (3,\ 5,\ 2)$. So, P_2, P_3, P_4 recover their own actual share point as similar way.

Now, any 3 participants say P_1, P_2, P_3 out of 4 want to reveal the secret using their shares. They can form the linear Eq. (3) as:

$$\begin{vmatrix} x_1 - 3 & x_2 - 5 & x_3 - 2 \\ 10 & 8 & 4 \\ 9 & 8 & 9 \end{vmatrix} = 0$$

i.e $7x_1 + x_2 + 8x_3 = 9\ (mod\ 11)$.

This is equivalent to Eq. (1.1). So, they are identical in graph and compute

$$\mathrm{U} = (9 - 1) / (7 + 9 + 1) \quad (\mathrm{mod}\,11).$$

Using (4)

$$\text{U} = 2 \pmod{11}.$$

Hence, reconstruct the secret

$$\begin{aligned} \text{S} &= \text{P} + \text{U}(\text{Q} - \text{P}) \\ &= (0, 5, 10) + 2\,(1, 9, 7) \quad (mod\ 11). \\ &= (2, 1, 2) \quad (mod\ 11). \end{aligned}$$

This is unique with assumption of dealer's secret point.

5 Security Analysis

Here, we illustrate the correctness and security of our proposed scheme.

If plotter E determines the equation of hyperplane (1), then he/she can get the secret point S. The plotter E might try to derive the dealer's own secret k or secret of each member P_j.

Proposition 1 *Shares or shadows of participants are secured.*

Proof

Case 1 Let k be not secure and k be derivable from public values α and g.

$$k = \log_g \alpha$$

Now, E can try to compute the share s_j of each P_j from public information β_j and their respective shadow $\hat{s}_j$ as

$$\begin{aligned} &\beta_j^{-k} \cdot \hat{s}_j \\ &= \beta_j^{(-k)} \beta \cdot s\beta_j.\beta_j^k \\ &= s_j \end{aligned}$$

Case 2: Let own secret p_j of each participant P_j be derivable from public values g and β_j.

$$p_j = \log_g \beta_j$$

Now, E computes $\alpha^{-p_j} \cdot \hat{s}_j$ and gets s_j.

But both of the above cases k and p_j are not derivable from $\alpha = g^k$ and $\beta_j = g^{p_j}$, respectively. Since $\alpha = g^k$ and $\beta_j = g^{p_j}$ are DLP problem which is NP-complete problem. So, k and p_j are secure. Therefore, s_j is secure. So, we conclude the proposition.

Proposition 2 *Any group of participants having less than t members cannot reconstruct the secret.*

Proof For the share of P_j is in the form $s_j\left(x_1^{(j)}, x_2^{(j)}, \ldots, x_t^{(j)}\right)$. The way to know the secret S is obtained in the following:

$$\begin{vmatrix} x_1 - x_1^{(1)} x_2 - x_2^{(1)} \ldots x_t - x_t^{(1)} \\ x_1^{(2)} - x_1^{(1)} x_2^{(2)} - x_2^{(1)} \ldots x_t^{(2)} - x_t^{(1)} \\ \ldots \\ \ldots \\ x_1^{(t)} - x_1^{(1)} x_2^{(t)} - x_2^{(1)} \ldots x_t^{(t)} - x_t^{(1)} \end{vmatrix} = 0$$

$$\text{i.e } \sum_{i=1}^{t} a_i x_i = c \ (\text{mod } p).$$

If any t or more P_j comes together with their respective share, choose any t among them and form the above equation. Since P_j's are distinct, their shares must be distinct and independent because of all are linearly independent points. So, the equation is either unique or equivalent to dealer's assumption. That is, they are identical in graph.

If participants are less than t, then they cannot form the above determinant because there have maximum $t-1$ row with t columns in each. So, they cannot form the equation. Therefore, any $t - 1$ shares cannot determine the secret S $\left(x_1^{(s)}, x_2^{(s)}, \ldots x_t^{(s)}\right)$. Hence proof the proposition.

Proposition 3 *The scheme is ideal secret sharing scheme.*

Proof Space of secrets and shares is same in same field $GF(p)$. Also, we have seen that *size* of the secret and share is also same. The *size* of the secret point S is $|S| = t$, and also size of each share point say s_j is $|s_j| = t$. So, the *information rate* is $|S|/|s_j| = 1$. Hence, the scheme is *ideal*.

6 Comparative Study

In this section, we compare our scheme with some existing relative schemes published earlier, like Shamir's [1], Blakley's [2], Tassa's [19], Hsu's [22], Li Bai's [11], Patil and Desmukh's [17] and Yuan et al.'s [24] secret sharing schemes. In the secret sharing schemes of [1, 2, 11, 17, 19, 22], distributed the actual share of each participant by the dealer. So the actual shares are not secured. Yuan et al. in [24] presented a dealer-free secret sharing scheme, where participants hold their own actual shares. So, actual shares are secured. But the scheme is not ideal and multi-secret. Table 1 shows that our proposed scheme is ideal, perfect, multi-secret and secrecy of share.

Table 1 Comparison table

Parameters	Shamir's scheme [1]	Blakley's secret sharing [2]	Hierarchical threshold SS scheme [19]	Ideal multi-secret sharing [22]	Li Bai's secret sharing [11]	Patil and Desmukh's scheme [17]	Yuan et al. [24]	Our proposed scheme
Generalisation of minimum threshold value	Generalise	Generalise	Generalise	Specialise (all participants should be present to reform the secret) (2,2) scheme	Generalise	Generalise	Generalise	Generalise
Ideal	Yes	Yes	No	Yes	Yes	Yes	No	Yes
Perfect	Yes	No	Yes	Yes	No	Yes	Yes	Yes
Multi-secret	No	No	No	Yes	No	Yes	No	Yes
Secrecy of share	No	No	No	No	No	No	Yes	Yes

7 Conclusion

We have described an ideal and perfect (t, n) threshold multi-secret sharing scheme using geometric approach in finite field. This scheme is perfect because information of $(t - 1)$ or lesser shares does not reveal the secret information. The scheme is ideal because the *size* requirement of the secret information is same that of the *size* of the participant's share. The each coordinates of secret point use as one of the secret key. In our scheme, no information is leaked about secret and shares also. The share of each participant is secured because the original share of the participant is not shared; the dummy or shadow of original share of each participant is shared here. In previously published schemes, like the scheme of Shamir, Blakley and Li Bai , where actual share of participant is directly shared by the dealer, the information is leaked. Whereas in our scheme, original shares of the participants are not shared with the participants and instead corresponding shadows are shared with them. So, our scheme is more secured than previously published schemes. Future work will include the proper implementation of the scheme. Also, the scheme can be modified to make the verifiable scheme.

References

1. Shamir, A.: How to share a secret. Commun. ACM **22**(11), 612–613 (1979)
2. Blakley, G.R.: Safeguarding cryptographic keys. Proc. AFIPS **48**, 313–317 (1979)
3. Blakley, G.R., Meadows, C.: Security of ramp schemes. In: Proceedings of CRYPTO '84, LNCS 196, Springer, pp. 242–269 (1985)
4. Mignotte, M.: How to share a secret. In: Proceedings of the Conference on Cryptography, pp. 371–375 (1982)
5. Harn, L.: Efficient sharing (broadcasting) of multiple secrets. IEE Proc. Comput. Digital Techniq. **142**(3), 237–240 (1995)

6. Hwang, R.J., Chang, C.C.: An on-line secret sharing scheme for multi secrets. J. Comput. Commun. **21**(13), 1170–1176 (1998)
7. Chien, H.Y., Jan, J.K., Tseng, Y.M.: A practical (t, n) multi-secret sharing scheme. In: IEICE Transformation of Fundamentals of Electronics, Communications and Computer, pp. 2762–2765 (2000)
8. Yang, C.C., Chang, T.Y., Hwang, M.S.: A (t, n) multi-secret sharing scheme. J. Appl. Math. Comput. **151**(2), 483–490 (2004)
9. Hwang, R.J., Lee, W.B., Chang, C.C.: A concept of designing cheater identification methods for secret sharing. J. Syst. Softw. **46**(1), 7–11 (1999)
10. Pang, L.J., Wang, Y.M.: A new (t, n) multi-secret sharing scheme based on Shamir's secret sharing. J. Appl. Math. Comput. **167**(2), 840–848 (2005)
11. Bai, L.: A Strong Ramp Secret Sharing Scheme Using Matrix Projection, Presented at the Second International Workshop on Trust. Security and Privacy for Ubiquitous Computing, Niagara-Falls, Buffalo, NY (2006)
12. Tang, C., Yao, Z.: A New (t, n)-threshold secret sharing scheme. In: International Conference on Advanced Computer Theory and Engineering, IEEE pp. 920–924 (2008)
13. Shi, R., Zhong, H.: A secret sharing scheme with the changeable threshold value. In: International Symposium on Information Engineering and Electronic Commerce, IEEE 2009 pp. 233–236
14. Chou,Y., Chih, H.L., Pao, C.L., Li, Y.C.: A (2, 3) threshold secret sharing scheme using Sudoku. In: Sixth International Conference on Intelligent Information Hiding and Multimedia Signal Processing. IEEE (2010)
15. Harn, L., Lin, C.: Strong (n, t, n) verifiable secret sharing scheme. J. Inform. Sci. **180**(16), 3059–3064 (2010)
16. Liu, Y.X., Harn, L., Yang, C.N., Zhang, Y.Q.: Efficient (n, t, n) secret sharing schemes. J. Syst. Softw. **85**(6), 1325–1332 (2012)
17. Patil, S., Desmukh, P.: A novel (t, n) threshold secret sharing using dot product of linearly independent vectors. Int. J. Adv. Res. Comput. Commun. Eng. **2**(7) (2013)
18. Guo, C., Chang, C.C.: A construction for secret sharing scheme with general access structure. J. Informat. Hiding Multimed. Signal Proc. **4**(1) (2013)
19. Tassa, T.: Hierarchical threshold secret sharing. J. Cryptol. **20**(2), 237–264 (2007)
20. Zhao, D., Peng, H., Wang, C., Yang, Y.: A secret sharing scheme with a short share realizing the (t, n) threshold and the adversary structure. J. Comput. Math. Appl. **64**, 611–615 (2012)
21. Su, J., Juan, T., Chuang Y.L., Li, M.J.: An online verifiable and detectable (t, n) multi-secret sharing scheme based on HS-TS. In: Proceedings of Fourth International Conference in Information Technology and Applications (ICITA2007), Harbin, China, EI , vol. I, pp. 56–59 January 15–18, 2007
22. Hsu, C., Cheng, Q., Tang, X., Zeng, B.: An ideal multi-secret sharing scheme based on MSP. J. Inf. Sci. **181**(7), 1403–1409 (2011)
23. Subba Rao, Y.V., Chakravarthy, B.: CRT based threshold multi secret sharing scheme. Int. J. Netw. Security, **16**(4), 249–255 (2014)
24. Yuan, L., Li, M., Guo, C., Choo, K.R., Ren,Y.: Novel threshold changeable secret sharing schemes based on polynomial interpolation. PLoS One **11**(10), e0165512 (2016). https://doi.org/10.1371/journal.pone.0165512

Solving Arithmetic Mathematical Word Problems: A Review and Recent Advancements

Sourav Mandal and Sudip Kumar Naskar

Abstract This paper studies the research problem of solving mathematical word problems (MWPs) and reviews the related research and methodologies. Word problems are any numerical problems written in natural languages like English, based on any subject domain (mathematics, physics, chemistry, biology, etc.), and MWPs relate to word problems in the mathematics domain. Solving MWPs has been a long-lasting open research problem in the field of natural language processing (NLP), machine learning (ML), and artificial intelligent (AI); however, unlike other research problems in NLP, ML, and AI, it has not made much progress. MWPs which can be easily solved by second-grade students can often pose serious challenges to MWP solvers due to its diverse problem types and varying degree of complexities. Understanding such problems written in natural language requires proper reasoning toward equation formation and answer generation. We restrict the review in this survey only to research on solving arithmetic word problems from elementary school level mathematics. We analyzed all the important methodologies proposed by researchers along with the datasets they used for training and evaluation. We studied the technical aspects of the system components and the algorithms relevant to their research along with the scopes, constraints, and limitations. This review paper also discusses the performance of different MWP solvers and provides observations on related datasets.

1 Introduction

A word problem is a mathematical exercise in mathematical education where significant background information related to problem is presented as texts written in natural language rather than in mathematical notation [41]. Arithmetic word prob-

S. Mandal (✉)
Haldia Institute of Technology, Haldia 721657, West Bengal, India
e-mail: sourav.mandal@hithaldia.in

S. K. Naskar
Jadavpur University, Kolkata 700032, India
e-mail: sudip.naskar@cse.jdvu.ac.in

P. Chandra et al. (eds.), *Information Technology and Applied Mathematics*,
Advances in Intelligent Systems and Computing 699,
https://doi.org/10.1007/978-981-10-7590-2_7

lems are part of elementary level school curriculum of mathematics subject and are generally introduced in the second or third grade. With arithmetic word problems, students gather basic linguistic knowledge along with familiarity of basic operations (addition, subtraction, multiplication, and division). Starting with basic arithmetic MWPs, students learn to solve MWPs with higher complexities (e.g., rate, probability, permutation, combination) as they move to the higher classes. How children understand such MWPs and solve them are one of the challenging research problems for the researchers of cognitive science and education psychology, enriched with sufficient literature.

Researchers of AI, machine learning, and NLP have long been targeting to build a computing system which can resemble the human cognitive perspective to solve MWPs in various ways; however, they have been able to solve only a few simple word problems of primary levels of school education. Some MWP solvers are available nowadays, but an efficient MWP solver which can solve a large number of diverse problems is still not available. Some simple low-complexity word problems are given below to understand arithmetic MWPs which the researchers have been trying to solve.

1. There are 2 pencils in the drawer. Tim placed 3 pencils in the drawer. How many pencils are now there in total? Answer: 5
2. There are 33 walnut trees and 20 maple trees currently in the park. Park workers will plant 44 walnut trees today. How many walnut trees will the park have when the workers are finished? Answer: 77
3. A packet of pen contains 6 pens. How many pens are in 5 packets? Answer: 30
4. There are 52 books on the shelf. They need to be shared equally among five tables in the classroom. How many books are left on the shelf? Answer: 2

Apparently, the word problems given above are very easy to solve for humans. However, when it comes to computers to solve them automatically from the problem texts, even the so-called easy problems become non-trivial. Challenges are immense and are in almost every step. Some of the challenges are given here.

- Diverse word problems with varying degree of complexity.
- Ambiguities due to conjunctions and co-references.
- Multiple verb/word sense (e.g., *book*).
- Multi-word problem (*Joan's high school* played …).
- Presence of missing and irrelevant information in the problem sentences.
- Temporal ambiguities relevant to the events occurring in the problems.
- Missing or sufficient background information required to solve the word problems.
- Accuracy/performance of the supporting machine learning and/or NLP tools/ techniques used.

How students become capable of solving such complex word problems by proper reasoning is still a topic of research in education psychology. A huge volume of research work have been carried by researchers to identify and represent various cognitive patterns of the students solving word problems; [15, 42] report such research efforts. The objective of the current review paper is not to analyze the works based on cognitive science/psychology methodologies, rather to analyze the research works which use computer systems for automatic understanding, reasoning, processing and generation of answers for MWPs. We categorize the work by their objectives and methodologies to demonstrate the historical evolution of the research works which started with the work of [2]. Mukherjee and Garain [30] reviewed major contributions in the field of automatic word problem solutions of various types till the year 2007 including [1]. They [30] analyzed arithmetic, rate and probability word problem solvers by demonstrating the methodologies of the then works as a sub-part of their review along with the descriptions of the works on solving word problems in physics, chemistry, and mechanics. Research in MWP has reinvigorated, and major advancements have been made since the year 2014. Researchers used various machine learning and NLP techniques and took the advantage of the advancements in these fields. We categorically explain all notable research works and contributions in this field till date in Sect. 2. Section 3 discusses all the standard datasets available for MWP research. In Sect. 4, we briefly discuss some relevant issues about some noteworthy methodologies described in Sect. 2 with comparison and the pros and cons followed by conclusions in Sect. 5.

2 Description of Systems and Research Methodologies

Systems and methodologies associated with MWP evolved with time since 1960s. Major advances were made with significant technological advancements in AI, NLP, and computer programming. However, the basic steps for solving MWPs are almost the same for majority of these works even if they followed different architectures and techniques. The common steps are given below.

1. Processing of the input MWP texts syntactically and/or semantically (NLP tasks) with preprocessing (if required) for extracting required information.
2. Understanding and/or reasoning and/or representing the processed MWP texts following the developed logics/algorithms/methodologies with the help of various methods like supervised, unsupervised learning.
3. Build/map/represent the desirable mathematical equation expression after reasoning.
4. Generate final answer from such equations after analyzing answer requirements from the question asked.

Early research on MWP used rule-based, propositional approaches which were limited to a fixed set of word problems and problem types. Later, the MWP solvers became more wide in scope and useful to a comparatively large number of problem types.

2.1 Early Stage (1964–2006)

In 1960s, research in AI and NLP was gaining pace. During this time, solving MWP was taken as an interesting research problem by researchers in cognitive science, psychology and education.

The Beginning (1964) Bobrow [2] first identified the MWP as computer solvable and developed a system named 'STUDENT'. He mentioned this as *algebra story problem* and programmed by 'LISP' language. 'STUDENT' was the first try of natural language communication with the computers and was able to solve a wide range of algebra word problems of specific structures like times, rate, sequence, percentage. They first constructed and used 'METEOR' which was consisting of sequence of rules each having string transformation with some control information. 'METEOR' with combination of 'LISP' implemented Wang's algorithm for providing theorem using propositional language (LISP 1.5 program manual, MIT press). 'STUDENT' system had two programs—'STUDENT' (for transforming algebra problems into equations) and 'REMEMBER' (stored all global information to solve a particular problem). He explained algorithm of his work with few examples on algebra word problems and how it was transformed to equations to generate answer. One of such example was "*If the number of customer Tom gets is twice the square of 20% of the number of advertisement he runs, and the number of advertisements he runs is 45, what is the number of Tom gets?*" The output given by the program for this problem was "*The number of customer Tom gets is 162*". In the first step, it substituted 'twice' with '2 times', 'square of' as 'square', and '%' as 'percent'. Then, the word of the whole problem was tagged by their functions in terms of transformation procedure by matching with dictionary entries like 'times' and 'percent' as 'op' or operator with their precedence level. The next step was to produce simple sentences (referred as kernel sentence) from the complex sentence by resolving conjunctions like 'if', ' but', etc, along with separation of question sentence. "The number of customer Tom gets" was identified as 'variable' which is the left side of the verb 'is' (equality) and for which answer has to be generated by processing the right side of the verb 'is' by forming and solving a set of parenthesizes equations given herewith

(EQUAL X00001(NUMBER OF CUSTOMER TOM GETS))(EQUAL (NUMBER OF ADVERTISEMENTS HE RUNS) 45)(EQUAL (NUMBER OF CUSTOMERS TOM GETS) (TIMES 2 (EXPT(TIMES 0.2 (NUMBER OF ADVERTISEMENTS HE RUNS)) 2))).

The system had very narrow semantic base, and the global information stored in 'REMEMBER' program was very small. It only could solve a tiny amount of algebra word problem of specific structures, but it was the first remarkable AI approach to solve word problem which was ahead of time.

The Progress (1981–1988) After this work of [2], next significant works were done in the decade of 1980. Few simulation programs were introduced which categorized the input problem types to solve them with propositional logic theory. The works simulated methodologies relevant to the then research on cognitive science approach to solve such problems which tried to represent human cognition of different levels for solving word problems. Riley et al. [31] introduced development of problem-solving skills in arithmetic with growing ages of students with growing conceptual and problem-solving knowledge based on the cognitive aspects. He categorized the input problem texts into four types—CHANGE, EQUALIZING, COMBINE, and COMPARE with specific sub-types. CHANGE, EQUALIZING represented action that causes increment or decrement of quantities, and COMBINE, COMPARE represented static relationship among quantities. The theoretical work was based on three system components—(1) comprehension, (2) mapping from conceptual relations to quantitative procedure, and (3) execution of procedure which dependent on two pre-defined knowledge schemata—'problem schemata' and 'action schemata'. 'Problem schemata' represented semantic network structures of the elements (quantities) with the relations (quantitative, temporal, logical for different problem categories) and 'action schemata' represented problem-solving procedures. Given the availability of learned schema in memory, a child maps the problem statements along with the quantities to the appropriate elements of the schematic structure. So successful solution of a word problem according to [31] was dependent on the availability of the conceptual representation (schemata) that was learned before [29].

At the same time, [3] proposed a model 'CHIPS' (Concrete Human-like Inferential Problem Solver)—a simulation program, which was also based on children psychology for solving word problem and the difficulty levels they faced. 'CHIPS' formulated the list structures, each representing a counter to represent a whole word problem as schema. Both the systems assumed that solution to word problem dependent on conceptual knowledge for problem representations. Kintsch and Greeno [16] developed a model that did both the text comprehension and problem solving based on arithmetic word problems. The model simulated construction of cognitive representations, information processing steps. The model analyzed processing requirements for *short-term memory (STM)* which is humanlike temporary memory, differing among types of problems. The model also extended the propositional strategies of [40] to represent a meaning of sentence "*John gave the book to Peter*" in terms of logical propositions, such as "*Gave to (a, b, c) & John = a, book = b and Peter = c*" and the problem-solving process. Although these works are all based on cognitive science for word problem solving, however, they made significant contributions along with problem type classifications (and problem sets) for further works which simulated their approach to make MWP as computer solvable. 'WORDPRO' [8], 'ARITHPRO' [7] were remarkable among some systems which were closely relevant to such human cognition science theories and understood specific input texts to generate the answers using computer.

Fletcher [8] wrote a computer program in 'Interlisp-D' to develop the system 'WORDPRO' which basically implemented the theory of comprehension and solution of [16] for third-grade arithmetic word problem. The system did not input any word problems text, instead it took input a set of propositions representing the textual meaning of a word problem referred as 'text base' working in the first level. In the second level, a 'problem model' drove the problem-solving process. He prepared a dataset of exactly 14 problems with the specific problem types—*CHANGE (Result*

Table 1 Propositional representation for change-(Result Unknown)

(Change-(Result Unknown),
(((PI,(EQUAL X JOE)) (P2,(PAST P3)),(P3,(HAVE X P4)),(P4,(3 MARBLE)))
((P5,(THEN P3 P7)),(P6,(EQUAL Y TOM)),(P7,GIVE Y X P9)),(P8,(PAST P7)) (P9,(5 MARBLE))) ((P1O,(HOWMANY MARBLE)),(PlI,(HAVE X P1O)),(P12,(NOW P11)))))

Unknown, Change Unknown, Start Unknown), COMBINE (superset unknown, subset unknown), COMPARE Difference Unknown, Compared Quantity Unknown, Referent Unknown) (cf. Table 1 of [8]*).* For example, "*Joe had three marbles. Then Tom gave him five marbles. How many marbles does Joe have now?*" was a *CHANGE (Result Unknown)* type. The system could solve all these 14 problems. The knowledge was represented by a set schema having three slots: an object slot (e.g., marbles), a quantity slot (e.g., 3), and a specification slot (distinguishes one set from another). Table 1 shows the 'text base' or propositional representation of that type for the given example. The 'problem model' was based on STM and a set of production rules (a set of conditions to be met applying the rules and a set of actions to be carried out) controlling the flow of information through STM to produce final answer (cf. Fig. 1 of [8]). The major observation/limitations of the input problems were that they had only the verbs 'has' and 'give', and question sentences were started with 'How many'. The system scope was very narrow and only solved a specific type of problems using propositional logic based on human cognitive approach and STM for solving.

Dellarosa [7] extended their own 'SOLUTION' system [6] and developed 'ARITHPRO' which was almost similar type of system [8] having many components were common like use of propositional frames to represent meaning of text [40] and STM concept to process intermediate data following the production rules. The major difference was its programming architecture includes a lexicon of knowledge (programmed using LOOPS), and a production system (programmed in INTERLISP). The lexicon consisted of possible words existing in simple word problems, arranged hierarchically following inheritance properties (cf. Fig. 1 of [7]). It triggered specific proposition production rules belonging to certain classes of lexicon. 'ARITHPRO' solved same 14 problems solved by 'WORDPRO' [8] and many more. As it used word classes in lexicon so any transfer or comparison word could be represented by same proposition schema to solve as per the sequence of production rules.

The Transitions (Up to 2006) After these researches of previous sections, no such remarkable works were published till the year 2007 by [1] with the system 'ROBUST'. But the researches based on cognitive science in this domain had been continuing all the time with new observations and methodologies. We will not discuss those works as those are out of scope of the present paper. Also during this time, major advances in the field of machine learning [27, 36] and NLP [13, 14, 24] had been occurring which definitely helped to introduce the same research problem with new solving methodologies. The technological progress in the field of machine learning and classifications techniques, syntactic and semantic analysis of natural language texts, various text categorization techniques [12, 43], etc., were used by many researchers to classify the various aspects of solving MWPs.

2.2 New Era (2007–2017)

The New Beginning with Complex MWPs (2007–2013) Bakman [1] developed advanced simulation program 'ROBUST' which could understand free-format multi-step arithmetic word problems with extraneous information, again based on propositional logic but for many verbs. He also categorized the 'CHANGE' verbs based on their role in MWPs into six distinct categories and developed schemas namely *Transfer-In-Ownership (e.g., to receive), Transfer-Out-Ownership (e.g., to lose), Transfer-In-Place (e.g., to bring), Transfer-Out-Place (e.g., to drag out), Creation (e.g., to build), and Termination (e.g., to eat)*. He used the example "*Ruth had 3 apples. She put 2 apples into a basket. How many apples are there in the basket now, if in the beginning there were 4 apples in the basket?*" to explain the methodologies. Previous works [7, 8] were based on single-step arithmetic problem (e.g., $A + B = C$). He introduced methodologies to solve multi-step MWPs. For the given example, there are many equations and the children have to write more than one step to solve such problem. The system also identified extraneous or irrelevant information ("*Ruth had 3 apples*") which had no role in the processing of answer generation procedure by analyzing the question asked for. Identification and removal of 'irrelevant sentences' or 'sentences with irrelevant information' throw huge challenges in solving MWPs. The given example belonged to 'Transfer-In-Place' schema, and the corresponding schema definition related to this was

"There were X objects in the place. Y objects were transferred into the place. There are Z objects in the place now."

Similar kinds of schemta were developed for other 'CHANGE' types and for 'COMPARE' and 'COMBINE' types of problems. The given example could be represented in the form of propositions using simple English like.

(1) Ruth had 3 apples. (2) Ruth put 2 apples into the basket.
(3) There are ? apples in the basket now. (4) There were 4 apples in the basket.

The schema instantiation belonged to "*Transfer-In-Place (initially 4, in 2, finally ?)*" that further initiated a addition operation between two quantities to generate answer '6'.

Another relevant work of that time was done by [4] to categorize MWP text in terms of two categories—'Multiplicative Compare (MC)' and 'Equal Group (EG)' problems using classification techniques based on Part of Speech (POS) Tagging. They also proposed a theory to identify 'irrelevant sentences' in the MWPs using classification techniques in their work of [5]. Matsuzaki et al. [26] developed an algorithm to represent a simple and logic-based architecture for solving MWPs. They proposed Discourse Representation Structure (DRS) for representing MWPs semantically. Then after many intermediate transformations in NLP pipeline in solver the final answer are generated.

The Major Advancements (2014–2017) The major advancements to solve various types of MWPs have been found in last 4 years along with published standard datasets. The MWP solver systems were more efficient than before with less number of constraints. Use of various NLP and machine learning techniques has made the understanding of input texts and processing of natural language datasets much easier. The current advancements on information extraction (IE), syntactic and semantic analysis, learning and classification techniques were effectively used in most of the research works, We studied all remarkable research works and grouped them in below types based on their system characteristics.

- Equation template based with statistical learning approach [19, 44].
- Equation expression tree representation and evaluation based [17, 32–35, 37].
- Entity, object representations with/without the state transition toward the mathematical formulation based (only addition–subtraction types) [9, 23, 28].
- Tag-based reasoning and knowledge inference based [20, 22].

Equation Template-Based Approaches: In recent years, the first and pioneer work was done by [19] who used template-based techniques (representation of information after IE) to store and process data and also prepared and tested on a standard dataset (514 problems). They used statistical learning approach to learn word problems with preset equation templates and equations which further processed to generate answers. Their research was related to three main fields of natural language processing.

1. **Semantic Interpretation**: To map natural language with the formal representation of meaning and multiple sentence reasoning.
2. **Information Extraction**: Related to template-based information extraction to identify event templates in MWP texts and fill the slots.
3. **Automatic Word Problem Solver**: Rule-based system to solve the problem from the mapped equation template on existing domain.

Their work was on fully supervised template-based information extraction and elementary level algebraic problem solution. Their objective was to map a given algebraic problem with the desired set of equations in equation templates with proper semantic meaning. They developed various equation templates, induced from the development set, with empty slots for number variables and unknown variables (aligned to nouns). They used supervised and semi-supervised learning method by gathering problems and solutions from www.Algebra.com. They used a equation template t, a number constant f, a number slot n and an unknown slot u referring noun or variable, a mathematical relation R on to two expressions (e.g., $n_1 u_1$), and a system template T referring a set of l equation templates $t_0, t_1, \ldots, t_l$ with the definition of alignment p as a set of pairs w, s where w was a token in the word problem and s is a slot instance in T. Then, an equation was constructed with the respective values for n, f, u according the appropriate template named *template instantiation*. By proper derivation and solving the generated equation, the final answer was produced. Figure 1 based on the research paper [19] shows an example of derivations

Derivation 1		
Word problem	An amusement park sells 2 kinds of tickets. Tickets for children cost \$ 1.50. Adult tickets cost \$ 4. On a certain day 278 people entered the park. On that same day the admission fees collected totaled \$ 792. How many children were admitted on that day? How many adults were admitted?	
Aligned template	$u_1^1 + u_2^1 - n_1 = 0$	$n_2 \times u_1^2 + n_3 \times u_2^2 - n_4 = 0$
Instantiated equations	$x + y - 278 = 0$	$1.5x + 4y - 792 = 0$
Answer	$x = 128$ $y = 150$	

Fig. 1 Two complete derivations for word problems. The derivation shows an alignment where two instances of the same slot are aligned to the same word (e.g., (u_1^1) and (u_1^2) both are aligned to noun Tickets and n is replaced real numbers [19]

into possible equations by proper template alignment. Kushman et al. [19] first proposed the techniques which were first of its kind. Eventually, the system was not performing good where more background knowledge and domain knowledge were required. However, they included MWPs with all basic operations like addition, subtraction, multiplication, division in their dataset and presented first learning result in the researches of solving MWPs.

Zhou et al. [44] work was also based on equation templates and mapping which was inspired by and could be recognized as the extension of the work of [19]. They used quadratic programming approach which simplified the template alignment procedure and derivation to equation and thus reduced hypothesis space (cf. Fig. 1 of [44]) with respect to [19]. Unlike [19], they did not assign the noun in unknown slot, and they only considered filling the number slot and described the relationship between numbers and unknowns by effective features. They demanded better accuracy in comparison with [19], but their system still could not resolve the word problems having complicated noun phrases and complex lexical features.

Equation Tree-Based Approaches: Some other researches [18, 32, 33, 37] followed an expression tree concept to represent equations on MWPs after synthesizing and processing and then solved. Koncel-Kedziorski et al. [18] introduced a MWP solver system ‘ALGES’, which represents multi-sentence MWPs into a set of possible equation trees (cf. Fig. 2). ‘ALGES’ only solved word problems based on single-variable equations. It also learned to map chunk of word problem text to related arithmetic operators and to choose the ‘best’ equation expression tree representing the problem. It covered all the basic operations like [19] and achieved more accuracy with respect to the dataset used by [9, 19]. Their research had mainly two components–*learning for every word problem-corresponding solution pair in training set and using inference with the learned model to generate solution for new word problems*. They used a compact representation for each node called a ‘Quantified Set (Qset)’ to build equation trees with quantities and their properties (e.g., ‘375 students’).

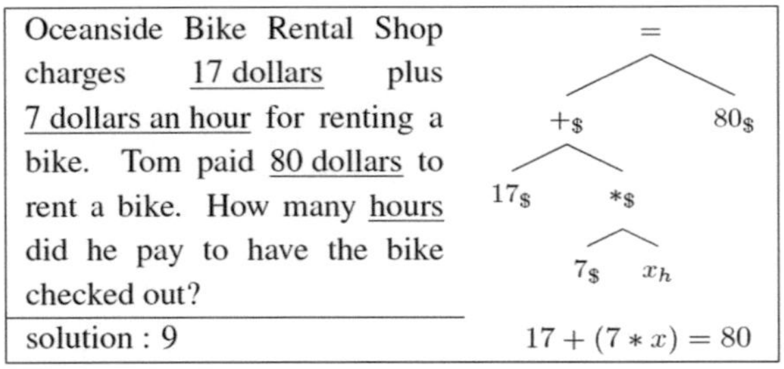

Fig. 2 Example word problem and possible equation tree representation and solution [18]

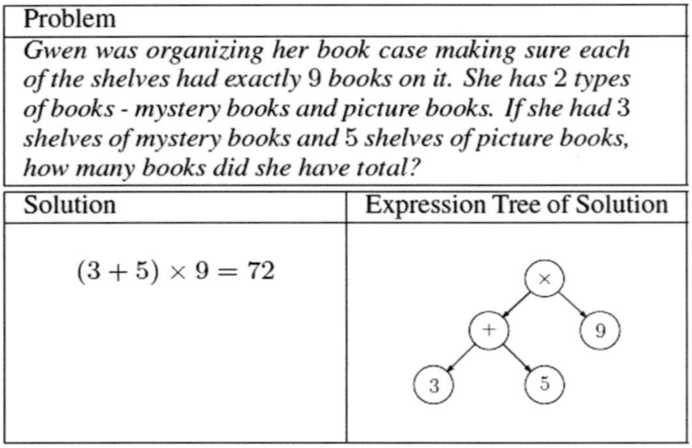

Fig. 3 An arithmetic word problem with the corresponding expression and expression tree [32]

A 'Qset' was a tuple consisting of *ent (quantity noun, e.g., students), qnt (number or quantity), adj (adjectives of ent), loc (location of ent), vrb (verb related to ent)*, etc. 'ALGES' extracts such 'Qset's from a problem texts and further combined and optimized based on 'Integer Linear Programming (ILP)' to generate a list of equation trees. They built a global discriminative model that scored and found the targeted 'best' equation trees based on the global problem structure. The final answer/solution generated from such equation tree, which was based on postfix notations, was done trivially.

Similar kind of equation expression tree concept has been used by Subhro Roy et al. [32–35] at Illinois University. They did good researches gradually to develop a MWP solver working on diverse word problems with publicly available datasets (Illinois(IL) dataset and commoncore dataset).[1] They developed a system using algorithmic approach that could handle multi-step and multi-operation arithmetic problems without depending on manual annotations or predefined tem-

[1] http://cogcomp.cs.illinois.edu/page/resource_view/98.

plates. They first developed various algorithms to identify quantity (of various types) with relevant information from a word problem [35]. They developed theory of expression tree [32] to represent a word problem and to evaluate. Their objective was to map a word problem to an arithmetic expression represented by an equation expression tree (cf. Fig. 3) that, when evaluated, provided the answer/solution. An expression tree τ was a binary tree. The leaf nodes represented quantities or numbers referred as $Q(n)$, and the internal non-leaf nodes represented any one operation among '+', '−', '∗', or '/' referred as $\odot(n)$ whose left child and right child were denoted as $lc(n)$ and $rc(n)$. The value associated with a node n represented as $Val(n)$, which could be recursively computed as follows

$$Val(n) = \begin{cases} Q(n), & \text{if } n \text{ is leaf node} \\ Val(lc(n)) \odot (n) Val(rc(n)), & \text{if } n \text{ is non-leaf node} \end{cases}$$

Finally, the $Val(n_{\text{root}})$ of the expression tree τ was evaluated as the answer of the equation or the word problem. They introduced and proved many theorems by identifying and defining various unique properties/features of such equation expression trees holding mathematical information which we do not discuss in this paper for the sake of page limits. But, the objective of such theorems on mathematical expression trees was to identify the operation between any two relevant quantities existing in the word problem along with their order of occurrence. They used two classifiers: one (using multi-class SVM) to predict operation between two quantities and another (relevance classifier using binary SVM) to identify irrelevant quantity in terms of final answer generation. They outperformed all previous system accuracies on existing datasets. They also introduced new standard datasets (illinois—562 problems and commoncore—600 problems) having more diverse problems with higher complexities. Their work was more generalized and broad with less dataset dependency. They extended their work to create a Web-based MWP solver[2] [33] which can solve a large number of word problems given by general user. They added a CFG perser with their existing MWP solver to manage queries asking for operations between numbers. They also recently developed a theory of 'unit dependency graph (UDG)' (e.g., unite refers 'flowers' in '8 flowers') to identify relationship and dependency between units of the quantities associated with given MWPs [34]. They used two classifiers—vertex (binary SVM) to detect 'rate' quantities and edge (multi-class SVM) classifier to detect properties of the edges (e.g., same unit, no relation). They integrated the concept of 'UDG' with their MWP solver [32] which reduced previous errors by 10%.

Another remarkable model for solving MWP in recent years was developed by Shi et al. [37]. Shi et al. [37] work was hugely different than above works as they worked on the dataset consisting of number word problem (a subset of MWPs) and developed an entirely new language to solve. They developed a system 'SigmaDolphin' which solved MWPs using semantic parsing and reasoning and a meaning representation language 'DOL' for the structured semantic representation of an MWP. First, a *semantic parser* transformed input texts into DOL representation tree (cf. Fig. 2 of

[2] Working demo available at http://cogcomp.cs.illinois.edu/page/demo_view/Math.

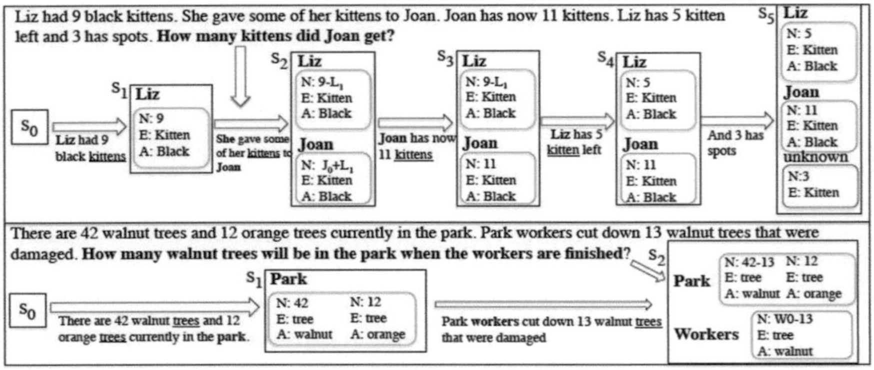

Fig. 4 System working principle with state transitions related to the work of [9]

[37]). Then, a *reasoning module* generated math equation from the representation. They prepared a dataset consisting of 1878 number word problems (e.g., "*Nine plus the sum of an even integer and its square is 3 raised to the power of 4. What is the number?*").

Entity and State Transition-Based Approaches: Hosseini et al. [9] developed a system 'ARIS' that solved only *addition–subtraction*-type MWPs. 'ARIS' represented whole MWP texts in a state representation (cf. Fig. 4) consisting of a set of entities like *'objects' (whose quantities are changing/considered, e.g., 'kitten' or 'tree'), 'attributes' (specifies the objects, e.g., 'black' or 'walnut'), their 'containers' (owner of the entities, e.g., 'Joan' or 'park'), quantities (associated cardinal values or unknown variables to be answered) and relations (among containers, entities, attributes and quantities)*. All these (Entities, Containers, Relations) were represented as world state as a tuple of $<E, C, R>$. Their work was based on verb categorization and state transitions of the world set (cf. Fig. 4) in the form of an algebraic operation of addition or subtraction related to verb category. They categorized the verbs into seven categories (cf. Table 1 of [9]) based on their operational meaning and features (similarity based, WordNet based, structural). For example, "*John lost 3 of the violet balloons.*", the verb *'lose'* was classified into *'negative'* category holding subtraction operation ('$-$'). 'ARIS' formed states $s_1, \ldots, s_T$ by updating quantities (if required) in every container according to verb categories learned given the sentence in the word problem. Finally, they analyzed the entities, container of the question sentence to select the appropriate container and its state progression to form the mathematical equation which was rule based (e.g., $9 - L1 = 5$ for the first example in Fig. 4). Solving the equation was then trivial to generate final answer. They compiled datasets containing 395 problems (namely AI[3] dataset) on addition–subtractions and tested on it to achieve remarkable accuracy. Hosseini et al. [9] did not mention how the addition of all relevant quantities for the question sentence having modifiers like 'together', 'in all' was represented with state diagram. Also,

[3]http://allenai.org/data.html.

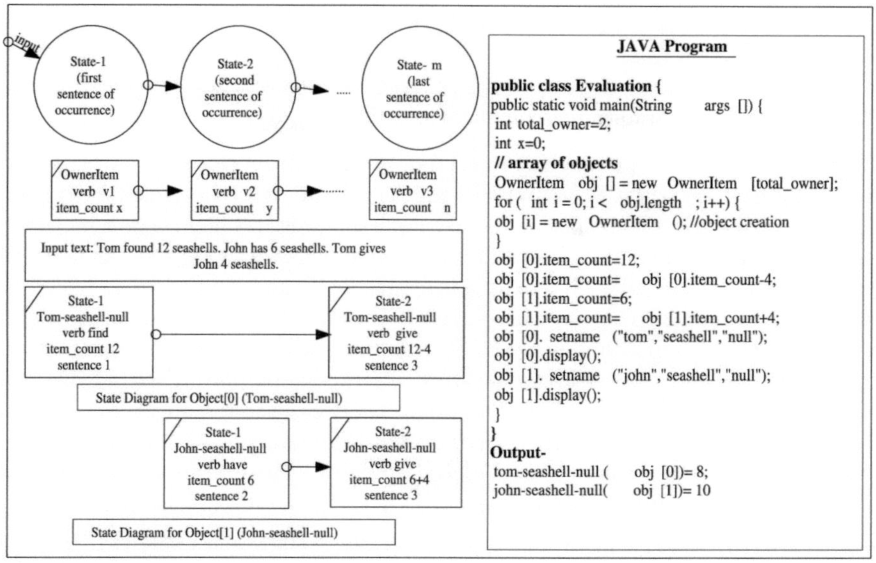

Fig. 5 System state transition and automatic program generation to the work of [23]

their answer generation part was specific to the datasets they used and verb categories whose operational meanings were based on single sense but exceptional cases could be easily thought.

The similar kind of representation of a MWP had been introduced recently by ourselves [23]. We used object-oriented modeling approach to model a word problem. The information extraction (IE) procedure was based on semantic role labeling techniques. The system identified each key entities and stored the relevant information in the predefined templates with the help of database management system (RDBMS). Then, the extracted information was modeled following the principle of object-oriented programming (OOP). We also categorized the verbs manually into five categories (*observation, increment, decrement, positive transfer, negative transfer*) based on their operational connotations. Then, state diagrams (cf. Fig. 5) were prepared for all the owner objects (similar to 'containers' of [9]). The answer generation part was very interesting as the system first generated a complete computer program from using *JAVA* programming language automatically from the system model representation and following state diagram which when executed, would generate the answer. The solution requirements are considered by analyzing the question sentence and some other factors based on which the appropriate state diagram(s) are identified for automatic program generation. We prepared our dataset by selected problems from the datasets of the work of [9]. Our work not exactly tried to solve MWPs rather tried to generate computer JAVA program automatically by modeling MWPs with OOP, and thus showed a new direction in solving them. Figure 5 shows the state representation and transitions with the generated program automatically.

Very recently, [28] developed a model to solve *AddSub* word problems of AI targeting problem formulation then solution. They categorized the whole problem set into three problem categories *Part-whole, Change, Compare* based on their structural features which lead to form mathematical equations.

Tag/Meaning Based Approaches: Recently, [10, 20–22] developed a role tag-based statistical meaning based MWP solver with understanding, reasoning and explanation. The system used Web interface to take input of a word problem texts which then went through a pipeline of processes. First a *Language Analyzer (LA)* module based on Stanford coreNLP suit [25] was used to syntactically and semantically obtain linguistic representation. Then, a *Solution Type Classifier (STC)* which was basically a SVM classifier identified among 6 possible solution types (e.g., *addition, subtraction, multiplication, sum, time-variant-quantities, or any combination*). Then, a two-stage methods namely *Logic Form Transformation* had been used to convert the linguistic representation into logic forms. First, the texts/ sentences were transformed into FOL predicates. For example "*Fred picked 36 roses*" was into FOL predicates as

"verb(v1,pick) & nsubj(v1,n1) & dobj(v1,n2) & head(n2,rose) & nummod(n2,36)"

where *v1, n1, n2* were the identifiers. Secondly, generic math facts related to the quantities were generated. FOL functions like "*quan(qid,unit,object)=number*" described facts about quantities (for the given example "*quan(q1,#,rose)=36*" was generated with auxiliary fact *verb(q1,pick)*–which were referred as **tags**). Next step was *Logic Inference*, which provided utilities to select desired facts on FOL instances and then determined required arithmetic operations on those selected facts. For example, Sum utility, *'Sum(function, condition)'*, returned the sum of the values of relevant FOL function instances. Finally, *Explanation Generation (EG)* explained the associated reasoning step (following the method developed by [10]) in the form of *Explanation Tree*, representing associated operation and quantities according to the solution type. Figure 6 shows the steps and intermediate information processing stated above of their MWP solver. They demanded best accuracy over AI and IL datasets so far using their approach. They used a simple example to explain their procedure. How they solved the complex problems of the datasets were not obvious from their explanation. Also, how generic their system was on the word problems out of the datasets as they used 6 specific solution types, was also not clear to the readers.

Those are the research works stated above in recent years who took challenges to solve such interesting problem and bit successful to some extents. In next section, we will discuss the dataset published so far explaining their characteristics and features.

3 Standard Datasets and Evaluations

Recently, Koncel-Kedziorski et al. [18] prepared a repository of word problems belonging to the standard datasets used in notable research works on MWP. Word problems can be of many types with many constraints and challenges which are beyond the capabilities of a single MWP solver as a whole. Therefore, the problems

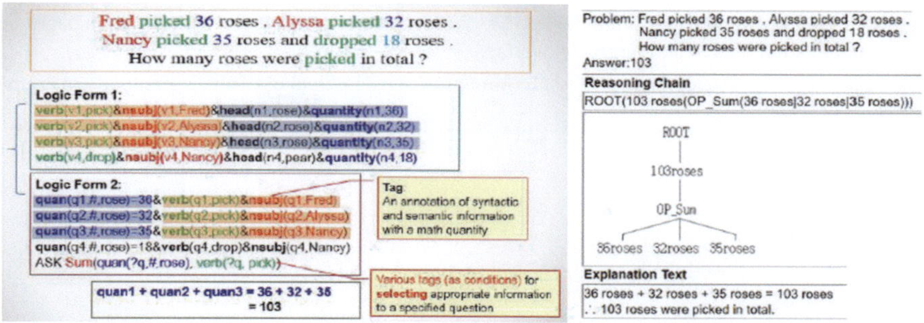

Fig. 6 The procedures of the MWP solver related to the work of [22]

are categorized into various types (relevant to the published works discussed earlier) to guide the researchers who want to take up any of these problems as future endeavors. Koncel-Kedziorski et al. [18] prepared an online repository—MAWPS[4] (MAth Word ProblemS) which consists of datasets of word problems with specific characteristics. MAWPS contains word problems without lexical and template overlaps as well as excludes problems with grammatical complications so as to provide unique problem types and to facilitate the addition of new problem types. Initially, they gathered the problems from the datasets of the published works given in Table 2, where *AddSub* refers to '*Addition–Subtraction*' problems, '*SingleOp*' relates to '*Single-Operation*' problems, etc. After that, they removed the ungrammatical problems which result in unwanted complications in solving MWPs. Then, they identified word problems with lexical and template overlaps which unnecessarily increase the count of similar types of problems in the datasets (cf. Sect. 2).

It is evident from Table 2 that the recent existing MWP solvers evaluated their performance on a small set of word problems limited to both scale and diversity. The reasons behind this may be unavailability of large training data or challenges in natural language understanding. The state-of-the-art MWP solvers' performance degrades on other datasets with different types of problems. Huang et al. [11] gathered a large-scale and diverse problem set consisting of 18,460 annotated elementary MWPs from community question answering (CQA) Web pages (math category of Yahoo!) by semiautomatically extracting problems, corresponding equation systems and answers to build *Dolphin18K*.[5] Their observation is very obvious —"the problem of automatic math word problem solving is still far from being solved". They showed some of the well known MWP solvers which achieved good results on small datasets and had worse performance on larger and diverse datasets like Dolphin18K.

[4]http://lang.ee.washington.edu/MAWPS.

[5]http://research.microsoft.com/en-us/projects/dolphin/.

Table 2 Standard datasets reported in [18] (MAWPS) consisting of MWPs with filtration based on ungrammaticality, lexical and template overlaps

Dataset	Problems	Grammar OK	Lexical overlap reduction (%)	Template overlap reduction (%)
AddSub published by [9]	395	357	22.8	11.3
SingleOp published by [35]	562	491	21.8	2.8
MultiArith published by [32]	600	526	17.0	10.9
SingleEq published by [18]	508	434	20.6	38.5
SimulEq-S published by [19]	514	437	21.7	76.8
SimulEq-L superset of SimulEq-S	1155	980	22.8	97.0

Upadhyay and Chang [39] proposed a method to evaluate performance of the existing MWP solvers using *derivations* which resembles equations. Upadhyay and Chang [38] also published a dataset 'DRAW' (DiveRse Algebra Word problem) with 1,000 annotated algebra word problems over a wide range of descriptions and equation systems.

4 Comparative Analysis and Discussions

In Sect. 2, we described all the relevant methodologies to solve various types of MWPs as given in Sect. 3. Major research works proposed different computer solvable methods and/or algorithms which can process and solve MWPs of specific structures. In the early stage of MWP research, starting from the work of [2], the works were rule-based and limited to a small number of datasets. The MWP solvers in the decade of 1980s were based on propositional logic and they mainly tried to implement the cognitive theories of word problem solving established by the researchers of psychology and cognitive science. The work of [7, 8] were almost similar and both of them solved word problems related to—change. Major advances in the field of NLP and machine learning (text classification) helped introduce newer approaches which solved diverse word problems in a large number. Use of classification techniques slowly replaced the rule-based approaches to solve such problem to a great extent. Identification of irrelevant quantities/information which adds more complexities in such word problems was first found in the work of [1]. Research on automatically solving MWPs has recently gained a momentum due to huge interest among the AI and NLP communities. Stronger methodologies have been introduced in recent years with supporting datasets (cf. Sect. 3). These methods (cf. Sect. 2) are not much similar" to each other except their use of various classification techniques required for various tasks. We described them in Sect. 2 with detailed analysis and grouped them in terms of their characteristics. The *expression tree based methodologies* introduced by Koncel-Kedziorski et al. [32] and Roy and Roth [17] are probably the most fruitful methods as they considered all operations with a strong IE task to process and represent such problems in equation expression trees for solving MWPs. Their algorithm is effective and can solve a huge number of MWPs with diverse complexities. The *template-based methods* are very limited to a fixed kind of word problems and their predefined equation templates. Representing all real-world word problems with equation templates, searching the most accurate equation template for a new word problem and representing the MWPs by the corresponding equations are very ambitious and not such effective in our opinion. The *entity (object) representation* and *state transition-based methodologies* are effective for addition–subtraction word problems of specific types. For this method to be successful, the word problems must have various entities and variables which could be easily represented by a world state and their transitions toward the solution. These methodologies were not tested on a large dataset with diverse problem types. *Tag-based approach* also works based on rules and knowledge inferences. Their answer generation technique was limited to the supported word problem; how the types could be extended is not obvious. A lot of researches are going on to develop benchmark dataset(s) with maximum diverse word problems on which different system methodologies can be evaluated.

5 Conclusions

In this paper, we described different research works on solving various types of mathematical word problems with reference to the datasets used by them for the evaluation of the system performance. We also discussed the important parts of their system components along with discussion on the pros and cons of the individual approaches. Automatically solving MWPs is a very challenging AI problem the basis of which are machine learning and natural language processing and/or understating. Solving such word problems has many applications like—*Mathematics tutoring, E-learning, calculations in commerce and finance domain, Question Answering in mathematics domain*. A complete MWP solver which can solve a large and diverse set of problems is yet to come. The system methodologies for solving word problems discussed in this paper are very discrete in nature differing across systems. Most of the researchers tried to solve the problem on relatively small-scale datasets and discontinued after a few publications. In recent years, research on MWP has taken an organized and goal oriented shape with the publication of standard datasets. Some research groups worldwide are making impressive developments with promising results on this topic. This paper will help researchers to understand the current status in MWP research in terms of different approaches, components, and datasets and make their future endeavors accordingly.

References

1. Bakman, Y.: Robust understanding of word problems with extraneous information (2007). arXiv: math/0701393
2. Bobrow, D.G.: Natural language input for a computer problem solving system (1964)
3. Briars, D.J., Larkin, J.H.: An integrated model of skill in solving elementary word problems. Cognit. Instr. **1**(3), 245–296 (1984)
4. Cetintas, S., Si, L., Xin, Y.P., Zhang, D., Park, J.Y.: Automatic text categorization of mathematical word problems. In: FLAIRS Conference (2009)
5. Cetintas, S., Si, L., Xin, Y.P., Zhang, D., Park, J.Y., Tzur, R.: A joint probabilistic classification model of relevant and irrelevant sentences in mathematical word problems. JEDM J. Educ. Data Min. **2**(1), 83–101 (2010)
6. Dellarosa, D.: Solution: a computer simulation of children's arithmetic word problem solving (Technical Report no. 148). University of Colorado, Institute of Cognitive Science, Boulder (1985)
7. Dellarosa, D.: A computer simulation of children's arithmetic word-problem solving. Behav. Res. Methods Instrum. Comput. **18**(2), 147–154 (1986)
8. Fletcher, C.R.: Understanding and solving arithmetic word problems: a computer simulation. Behav. Res. Methods **17**(5), 565–571 (1985)
9. Hosseini, M.J., Hajishirzi, H., Etzioni, O., Kushman, N.: Learning to solve arithmetic word problems with verb categorization. In: Proceedings of the 2014 Conference on Empirical Methods in Natural Language Processing, EMNLP 2014, 25–29 October 2014, Doha, Qatar, A meeting of SIGDAT, a Special Interest Group of the ACL, pp. 523–533 (2014). http://aclweb.org/anthology/D/D14/D14-1058.pdf
10. Huang, C., Lin, Y., Su, K.: Explanation generation for a math word problem solver. IJCLCLP **20**(2) (2015). http://www.aclclp.org.tw/clclp/v20n2/v20n2a2.pdf
11. Huang, D., Shi, S., Lin, C., Yin, J., Ma, W.: How well do computers solve math word problems? Large-scale dataset construction and evaluation. In: Proceedings of the 54th Annual Meeting of the Association for Computational Linguistics, ACL 2016, 7–12 August, 2016, Berlin, Germany, Vol. 1: Long Papers (2016). http://aclweb.org/anthology/P/P16/P16-1084.pdf
12. Joachims, T.: Text categorization with support vector machines: Learning with many relevant features. Machine learning: ECML-98 pp. 137–142 (1998)
13. Jurafsky, D.: Speech and Language Processing. Pearson Education India (2000)
14. Jurafsky, D., Martin, J.H.: Speech and Language Processing, vol. 3. Pearson (2014)
15. Kingsdorf, S., Krawec, J.: A broad look at the literature on math word problem-solving interventions for third graders. Cogent Educ. 3(1), 135770 (2016). https://doi.org/10.1080/2331186X.2015.1135770
16. Kintsch, W., Greeno, J.G.: Understanding and solving word arithmetic problems. Psychol. Rev. **92**(1), 109 (1985)
17. Koncel-Kedziorski, R., Hajishirzi, H., Sabharwal, A., Etzioni, O., Ang, S.D.: Parsing algebraic word problems into equations. TACL **3**, 585–597 (2015). https://tacl2013.cs.columbia.edu/ojs/index.php/tacl/article/view/692
18. Koncel-Kedziorski, R., Roy, S., Amini, A., Kushman, N., Hajishirzi, H.: MAWPS: A math word problem repository. In: NAACL HLT 2016, The 2016 Conference of the North American Chapter of the Association for Computational Linguistics: Human Language Technologies, San Diego California, USA, 12–17 June, 2016, pp. 1152–1157 (2016). http://aclweb.org/anthology/N/N16/N16-1136.pdf
19. Kushman, N., Zettlemoyer, L., Barzilay, R., Artzi, Y.: Learning to automatically solve algebra word problems. In: Proceedings of the 52nd Annual Meeting of the Association for Computational Linguistics, ACL 2014, 22–27 June 2014, Baltimore, MD, USA, vol. 1, Long Papers, pp. 271–281 (2014)
20. Liang, C., Hsu, K., Huang, C., Li, C., Miao, S., Su, K.: A tag-based english math word problem solver with understanding, reasoning and explanation. In: Proceedings of the Demonstrations

Session, NAACL HLT 2016, The 2016 Conference of the North American Chapter of the Association for Computational Linguistics: Human Language Technologies, San Diego California, USA, 12–17 June 2016, pp. 67–71 (2016)
21. Liang, C., Hsu, K., Huang, C., Li, C., Miao, S., Su, K.: A tag-based statistical english math word problem solver with understanding, reasoning and explanation. In: Proceedings of the Twenty-Fifth International Joint Conference on Artificial Intelligence, IJCAI 2016, New York, NY, USA, 9–15 July 2016, pp. 4254–4255 (2016). http://www.ijcai.org/Abstract/16/647
22. Liang, C., Tsai, S., Chang, T., Lin, Y., Su, K.: A meaning-based english math word problem solver with understanding, reasoning and explanation. In: COLING 2016, 26th International Conference on Computational Linguistics, Proceedings of the Conference System Demonstrations, 11–16 December 2016, Osaka, Japan, pp. 151–155 (2016). http://aclweb.org/anthology/C/C16/C16-2032.pdf
23. Mandal, S., Naskar, S.K.: Towards generating object-oriented programs automatically from natural language texts for solving mathematical word problems. In: Natural Language Processing and Information Systems-22nd International Conference on Applications of Natural Language to Information Systems, NLDB 2017, Liège, Belgium, 21–23 June, 2017, Proceedings, pp. 222–226 (2017). https://doi.org/10.1007/978-3-319-59569-6_26
24. Manning, C.D., Schütze, H., et al.: Foundations of Statistical Natural Language Processing, vol. 999. MIT Press (1999)
25. Manning, C.D., Surdeanu, M., Bauer, J., Finkel, J.R., Bethard, S., McClosky, D.: The Stanford CORENLP natural language processing toolkit. In: Proceedings of the 52nd Annual Meeting of the Association for Computational Linguistics, ACL 2014, 22–27 June 2014, Baltimore, MD, USA, System Demonstrations, pp. 55–60 (2014). http://aclweb.org/anthology/P/P14/P14-5010.pdf
26. Matsuzaki, T., Iwane, H., Anai, H., Arai, N.H.: The complexity of math problems-linguistic, or computational? In: IJCNLP, pp. 73–81 (2013)
27. Mitchell, T.M., et al.: Machine learning. WCB (1997)
28. Mitra, A., Baral, C.: Learning to use formulas to solve simple arithmetic problems. In: Proceedings of the 54th Annual Meeting of the Association for Computational Linguistics, ACL 2016, 7–12 August 2016, Berlin, Germany, vol. 1: Long Papers (2016). http://aclweb.org/anthology/P/P16/P16-1202.pdf
29. Morales, R.V., Shute, V.J., Pellegrino, J.W.: Developmental differences in understanding and solving simple mathematics word problems. Cognit. Instr. **2**(1), 41–57 (1985)
30. Mukherjee, A., Garain, U.: A review of methods for automatic understanding of natural language mathematical problems. Artif. Intell. Rev. **29**(2), 93–122 (2008). https://doi.org/10.1007/s10462-009-9110-0
31. Riley, M.S., et al.: Development of children's problem-solving ability in arithmetic (1984)
32. Roy, S., Roth, D.: Solving general arithmetic word problems. In: Proceedings of the 2015 Conference on Empirical Methods in Natural Language Processing, EMNLP 2015, Lisbon, Portugal, 17–21 September, 2015, pp. 1743–1752 (2015). http://aclweb.org/anthology/D/D15/D15-1202.pdf
33. Roy, S., Roth, D.: Illinois math solver: math reasoning on the web. In: Proceedings of the Demonstrations Session, NAACL HLT 2016, The 2016 Conference of the North American Chapter of the Association for Computational Linguistics: Human Language Technologies, San Diego California, USA, 12–17 June 2016, pp. 52–56 (2016). http://aclweb.org/anthology/N/N16/N16-3011.pdf
34. Roy, S., Roth, D.: Unit dependency graph and its application to arithmetic word problem solving. In: Proceedings of the Thirty-First AAAI Conference on Artificial Intelligence, 4–9 February 2017, San Francisco, California, USA, pp. 3082–3088 (2017). http://aaai.org/ocs/index.php/AAAI/AAAI17/paper/view/14764
35. Roy, S., Vieira, T., Roth, D.: Reasoning about quantities in natural language. TACL **3**, 1–13 (2015). https://tacl2013.cs.columbia.edu/ojs/index.php/tacl/article/view/452
36. Sebastiani, F.: Machine learning in automated text categorization. ACM Comput. Surv. (CSUR) **34**(1), 1–47 (2002)

37. Shi, S., Wang, Y., Lin, C., Liu, X., Rui, Y.: Automatically solving number word problems by semantic parsing and reasoning. In: Proceedings of the 2015 Conference on Empirical Methods in Natural Language Processing, EMNLP 2015, Lisbon, Portugal, 17–21 September 2015. pp. 1132–1142 (2015). http://aclweb.org/anthology/D/D15/D15-1135.pdf
38. Upadhyay, S., Chang, M.W.: Draw: A challenging and diverse algebra word problem set. Technical Report, Number MSR-TR-2015-78 (2015)
39. Upadhyay, S., Chang, M.: Annotating derivations: A new evaluation strategy and dataset for algebra word problems (2016), arXiv: abs/1609.07197
40. Van Dijk, T.A., Kintsch, W., Van Dijk, T.A.: Strategies of Discourse Comprehension. Academic Press, New York (1983)
41. Verschaffel, L., Greer, B., De Corte, E.: Making sense of word problems. Lisse Swets and Zeitlinger (2000)
42. Wang, A.Y., Fuchs, L.S., Fuchs, D.: Cognitive and linguistic predictors of mathematical word problems with and without irrelevant information. Learn. Individ. Differ. **52**, 79–87 (2016)
43. Yang, Y., Liu, X.: A re-examination of text categorization methods. In: Proceedings of the 22nd Annual International ACM SIGIR Conference on Research and Development in Information Retrieval, pp. 42–49. ACM (1999)
44. Zhou, L., Dai, S., Chen, L.: Learn to solve algebra word problems using quadratic programming. In: Proceedings of the 2015 Conference on Empirical Methods in Natural Language Processing, EMNLP 2015, Lisbon, Portugal, 17–21 September 2015, pp. 817–822 (2015)

Pose-Invariant Hand Geometry for Human Identification Using Feature Weighted *k*-NN Classifier

Asish Bera, Debotosh Bhattacharjee and Mita Nasipuri

Abstract Hand biometrics is globally deployed for automated human identification based on the discriminative geometric characteristics of hand. Advancements in hand biometric technologies are accomplished over several decades. The key objectives of this paper are two-fold. Firstly, it presents a comprehensive study on the state-of-the-art methods based on the hand images collected in an unconstraint environment. Secondly, a pose-invariant hand geometry system is excogitated. The experiments are conducted with the weighted geometric features computed from the fingers. The feature weighted *k*-nearest neighbor (*fwk*-NN) classifier is applied on the right- and left-hand images of the 500 subjects of the Bosphorus database for performance evaluation. The classification accuracy of 97% has been achieved for both of the hands using the *fwk*-NN classifier. Equal error rates (EER) of 5.94% and 6.08% are achieved for the right- and left-hand 500 subjects, respectively.

Keywords Biometrics · Feature weight · Hand · Pose-invariant · Weighted *k*-NN

1 Introduction

Secured applications are extremely indispensable to protect the assets and properties from unauthorized access. Biometrics provides reliable and secure human authentication to access the commercial, government, surveillance, Internet of things (IoT),

A. Bera (✉)
Haldia Institute of Technology, Haldia 721657, WB, India
e-mail: asish.bera@gmail.com

D. Bhattacharjee · M. Nasipuri
Jadavpur University, Kolkata 700032, WB, India
e-mail: debootsh@ieee.org

M. Nasipuri
e-mail: mitanasipuri@gmail.com

P. Chandra et al. (eds.), *Information Technology and Applied Mathematics*,
Advances in Intelligent Systems and Computing 699,
https://doi.org/10.1007/978-981-10-7590-2_8

and forensic applications [1, 2]. Biometric recognition is an active area of research in the fields of pattern recognition and machine learning since the last five decades [3]. It is a viable tool used for automated personalization and real-time systems in a restricted-access environment than the knowledge-based (password) or token-based (PIN) authentication schemes. In recent years, various electronic devices are connected through the IoT for which biometrics is a secure authentication tool [2]. Even, the face, iris, and other modal recognition are deployed in the smartphone.

Hand geometry is regarded as one of the oldest physiological biometric traits. The *Identimat* was the oldest device that is reliant on hand geometry since the earliest patent of Miller in 1971 [4]. An approach for biometric security in a transponder-reader system based on hand geometry using the radio frequency identification (RFID) technology is patented in 2011 [5]. Hand biometry is useful for automated attendance system mainly in the nuclear plant, border control of different nations, and others. It is advantageous in those circumstances when a lower cost of the 2-D sensors, smaller feature template (a few bytes) sizes, and easier accessibility with user-friendliness are preferred [6, 7]. Applications of biometric-enabled services are also growing in developing nations, like India. The Unique Identification Authority of India (UIDAI) has implemented a multimodal system with additional demographic information, namely the *Aadhaar* card since 2009 [3]. It is one of the largest national biometric systems in the world.

Various geometric- and shape-based descriptors are computed from the fingers and palm in the context of pose-invariant hand biometrics [8]. The Fourier descriptors [9], wavelet descriptors [10], and scale-invariant feature transform (SIFT) represent shape-based features [11]. Geometric measurement is simpler to compute than silhouette-based feature. The shape-oriented feature is attractive due to its invariance properties such as translation, rotation, and scaling. Researches on hand biometrics using contact-less hand images have been significantly improved in recent times. This paper presents a hand geometry system with an emphasis on the progress of state-of-the-art methods of this trait, particularly, after the study presented by Duta [8]. The present approach is experimented using a variation of the weighted k-NN classifier [12] over the subjects of the Bosphorus database [7, 13]. The key objectives are as follows:

i. A pose-invariant hand normalization approach has been implemented and tested with the geometric features on the 500 subjects of the Bosphorus hand database.
ii. The feature weighted k-NN (*fwk*-NN) classification algorithm is devised in the context of hand biometrics.
iii. Improved performances have been achieved using the *fwk*-NN classifier through weight assignment on the feature space.

The rest of this paper is organized as follows: Sect. 2 describes the related works. The key steps for pose-invariant hand image normalization are presented in Sect. 3. The *fwk*-NN classification algorithm is described in Sect. 4. Performances of the present work are presented in Sect. 5. Finally, the conclusion is drawn in Sect. 6.

2 Related Work and Challenging Issues

Mainly, the 2-D and 3-D hand images have been experimented in the literature. Recently, the infrared (IR) [14, 15], hyperspectral [16], and synthetic hand images [17] are also emphasized. Moreover, the quality (dpi) of images is an important factor during hand segmentation at preprocessing stage. The resolution and dimension of the images alter due to various sensing devices such as the scanner or camera. Advanced sensors such as the infrared (IR) and hyperspectral images are robust against the spoof-attacks and provide better security against the environmental hazards. However, these sensors are costly and yet not deployed widely in traditional access control systems. Moreover, all the features are not discriminative enough while included in the feature vector for experiments. Thus, selection of relevant features plays an imperative role in performance evaluation. Information theory metric such as mutual information (MI) is applied for discriminative feature selection [18]. Many novel approaches are designed and implemented using pose-invariant hand images. Notably, it is beyond the scope to summarize all the methods. Only a few of the approaches are provided in Table 1.

Tip-valley points are used as reference points for geometric feature measurement. The size of feature template and computational complexity of shape descriptors are higher than geometric features. Dimensionality reduction technique for selecting discriminative features such as the principal component analysis (PCA) is generally followed in the literature. Some geometric and shape descriptors are listed in Table 2.

Most of the available hand databases are proprietary. A few are freely available for research. Mainly, the Bosphorus, IIT Delhi, CASIA, and GPDS are some popular and freely available databases [18, 19]. Recently, some more databases are created and few of which are mentioned in Table 3.

It has been studied that hand biometrics is still facing several challenges. Some important issues in pose-invariant hand geometry are as follows:

(a) Freeness and flexibility of hand pose cause angular variation between the fingers which may affect significant variation in tip-valley localization. These points are located erroneously due to arbitrary finger alignment.
(b) Any hand accessory such as a ring, wristwatch, or clothing occlusion can cause incomplete contour extraction which in turn causes a discontinuity in the hand contour. It may also lead to a higher intra-class variation at different imaging session(s) when the time lapse is sufficiently large. Moreover, the finger nail is an important factor that should be eliminated during preprocessing.

Table 1 Some state-of-the-art methods on pose-invariant hand biometrics

Authors	Feature definition	Experimental description
Dutağaci et al. [7]	Shape-based and appearance-based ICA features of both hands Maximum feature space is 600	Bosphorus 800(3) subjects; feature and score-level fusion EER: 0.33% for 400 genuine subjects Identification: 99.92%
Kanhangad et al. [6]	25 geometric features, Comp Code from a fingerprint, and Surface Code from palm print are defined	177(10) subjects; score-level fusion of 2-D and 3-D hand geometry features EER: 2.3%. Identification: 98.88%
Wang and Chung [14]	34 geometric features are extracted using thermal images Extension theory is applied for recognition	30(10) subjects; the neural network is used for recognition EER: not reported. Identification: 92%
Luque-Baena et al. [18]	Geometric- and shape-based features are defined On average 50 relevant features are selected using the (GA) and MI	GPDS (144 subj.), IITD (137 subjects), CASIA (100 subjects) databases are used. EERs: CASIA 4.64%; IITD 4.51%
Kang et al. [9]	A shape-based method using cumulative angular function (CAF) and Fourier descriptors (FDs) are extracted	Bosphorus database 638(3) subjects with left-hand images at score-level fusion EER: 3.69%. Identification: not reported
Sharma et al. [10]	1-D wavelet at level-5 using Daubechies-1 wavelet filter. 50 wavelet coefficients from each of the distance map and orientation map and 21 geometric features are defined	JUET (50 subjects) and IITD (240 subjects); two levels of matching score-level fusion EERs: IITD 0.52%; JUET: 0.31%. Identification: not reported

Table 2 Hand biometric features

Feature	Name
Geometric	(a) Length, (b) width, (c) area, (d) perimeter, (e) aspect ratio, (f) compactness, (g) solidity, (h) rectangularity, (i) inter-finger angles, (j) distances from centroid, (k) major-axis and minor-axis lengths of the fitted ellipse, (l) equivalent diameter of a circle with the same area
Shape-based	(a) Fourier descriptor, (b) wavelet descriptor, (c) HU moment invariants, (d) Zernike moments (h order), (e) shape coding using discrete hidden Markov models (DHMM), (f) independent component analysis (ICA), (g) appearance-based ICA, (h) distance transform (DT), (i) angular radial transform (ART), (j) coherent distance shape context (CDSC), (k) scale-invariant feature transform (SIFT)
Others	(a) Palm texture, (b) finger texture, (c) 3-D palm print (Surface Code), (d) feature-level fusion (CCA), (e) hyperspectral features

Table 3 Specification of some existing hand databases

Authors, Year	Database description
Tecnocampus Hand Image Database [15]	Right, 100 (5 sessions × 2 acquisitions × 3 spectral ranges × 2 sides (palm and dorsal images)) = 6000. VIS & TIR: thermal, Testo 882-3 camera. NIR images by a Logitech Webcam, changing optical filter using Kodak filters for IR. VIS and NIR: 640 × 480; TIR: 320 × 240 pixels
Guo et al. [20]	Right, 100 (60) = 6000. Commercial Webcam, IR filter, infrared illumination devices are used; IR image is 640 × 480 pixels
Kanhangad et al. [6]	Right, 177 (2 sessions × 5 samples per session × 2 types (3-D and 2-D)) = 3540 Commercial 3-D digitizer: Vivid 910 3-D digitizer 2D + 3D; 640 × 480 pixels
CASIIM hand database [19]	Left, 200 (2 sessions × 10 samples per session) = 4000. Digital CCD camera: Cannon Powershot SX110 IS; visible image 1000 × 1000 pixels
Synthetic [17]	Right, 100,000 (5) = 500,000. Parameters of the synthesizer are calibrated using the GPDS Hand database, which includes 10 images from of 144 users-captured using a digital scanner (150 dpi)
Bosphorus [7]	Both, 918 (3) = 5508. HP Scanjet scanner, 383 × 526 pixels (45 dpi)
Mid-range IR-GPDS database [21]	Right, 100 (10) = 1000.1470 nm (mid-range infrared) by a camera sensor XENICS XEVA 1.7–320. 320 × 256 pixels, (150 dpi)
SWIR hyperspectral hand [16]	Right, 154 (10 samples × 350 images per sample, palm and dorsum) = 539,000. SWIR Camera 256 bands between 900 nm and 1600 nm; 320 × 256 pixels

(c) Hand shape of a person changes with time and age. Thus, template adaptation is essential for deployment.
(d) The performance of a hand geometry system may degrade drastically when only a few features are used for identification of a large-scale population. A good feature selection algorithm may also be used for selecting discriminative features for performance improvement.
(e) Performances of the existing pose-invariant systems are not evaluated on any benchmark dataset. Hence, it is very difficult to compare pros and cons of the state-of-the-art methods directly.

Most of the works have implemented for human authentication in commercial and government purposes. In addition, hand biometrics in digital forensics can be explored. Pose-invariant finger geometry can also be deployed on the mobile devices

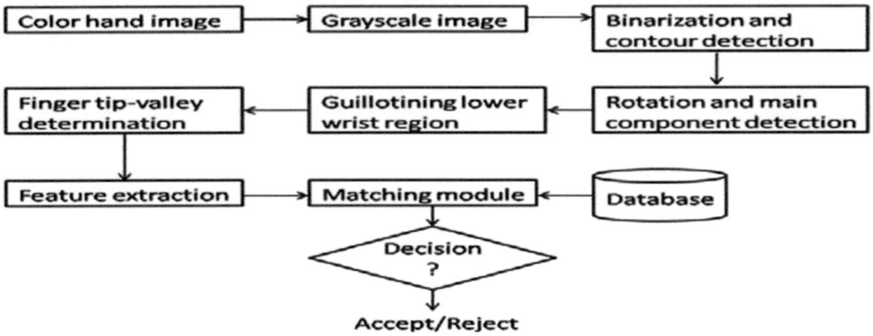

Fig. 1 Generalized hand biometric system

where the computation time is strictly limited to within a few seconds. Thus, for advanced communication and accessing devices through the IoT, finger biometrics can be thought as an alternative solution. Thus, it is a matter of concern that the off-the-shelf computer-based authentication approaches using hand biometrics can be deployed on the advanced hardware or mobile devices.

3 Pose-Invariant Hand Biometric Framework

Unconstrained hand biometric systems are developed to provide user flexibility to place the hand during image acquisition and preferred due to the absence of guiding pegs. A vignette of pose normalization method is described here.

3.1 Pose-Invariant Hand Image Normalization

A robust preprocessing technique, also known as hand image normalization in a contact-free environment, is essential to minimize intra-class variation before feature extraction. A method for pose-invariant hand image normalization is described in [22]. In this paper, a contact-less hand biometric system is presented which consists of successively sub-steps, as shown in Fig. 1.

The acquired hand images include actual hand shape with a darker background. The first step is to remove the background debris and segment the foreground hand. A color hand image is converted into a grayscale image using thresholding by the *Otsu's* method. It is a popular method and applied in several other works [11, 18]. Noise removal is an important task for hand segmentation. Noise can be associated due to several reasons such as the dirt artifacts and hand accessories. Thus, a suitable filter is required to remove associated noise. Here, a median filter is preferably applied. Then, the grayscale image is converted into the binary image. The binary hand may contain some holes and disconnected blobs. Next, the area-based filtering method is applied to remove unnecessary and smaller components for segmenting the actual hand region, also known as the region of interest (ROI). Next, hand contour is detected using an edge detector from the binary image using the *Canny's* edge detection algorithm. The *Sobel's* edge detector is also commonly used in the literature. Sometimes, hand contour is disconnected and fragmented into multiple components. In such cases, morphological operators are used to smoothing the edge and to remove spurious pixels. The largest connected component is selected as the main component which is used for further processing next.

The hand images are acquired at different global postures and angular variations of fingers. In pose-invariant imagery scenario, a uniform orientation of the main contour is essential for accurate feature computation. Angular variation of the hand can be normalized by rotating the contour as required to make it perpendicular with the X-axis with respect to its centroid. Ellipse fitting method, i.e., fitting of an ellipse over the hand contour, is a good method for rotation. This method is based on the rotation of the contour toward the larger eigenvector. An ellipse is fitted so that its major axis passes through the centroid of hand contour. Rotation is defined as the angle (θ) between the Y-axis and the major axis of fitted ellipse. The centroid is defined as:

$$(x_c, y_c) = \left(m_{1,0}/m_{0,0}, m_{0,1}/m_{0,0}\right) \tag{1}$$

The angle of rotation is defined as:

$$\theta = 0.5 \tan^{-1}\left(\frac{2\mathrm{m}_{1,1}}{\mathrm{m}_{2,0} - \mathrm{m}_{0,2}}\right), \text{ where } m_{ij} \text{ is the moment of the image.} \tag{2}$$

Rotation of the hand contour may cause a change in pixel intensities mainly at the boundary region. The bilinear interpolation is applied to stave off shape deviation. Another issue may also arise at the wrist area during hand prototyping. The lower portion of the wrist may contain shape irregularities because of hand accessories such as a ring, bracelet, or clothing variation at different sessions of the same person. This results in significant intra-class variation during feature computation. A smoothing process is followed to eliminate wrist irregularities, named as guillotine [13]. A fixed vertical distance of one hundred pixels from the centroid to the palm area has been neglected to eliminate wrist area [9]. Here, a reference line is considered to remove the wrist which depends on the length of the hand. About 15–20% from the below of the hand region is preferred here to locate two endpoints of the reference line with equal distance from the centroid. Next, the lower palm area below the reference line is eliminated which results in the normalized hand silhouette.

Localization of the tip-valley points on fingers is next task. These extreme points can be determined by scanning the contour pixels at the boundary which are located at the maximum Euclidean distance from the reference point. The midpoint of reference line can be delineated as the reference point. The extreme top pixel of every finger,

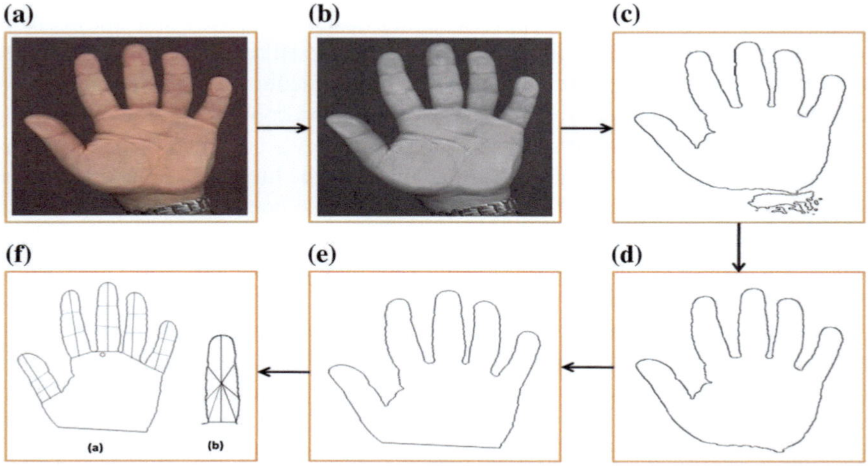

Fig. 2 Different steps of normalization for pose-invariant hand biometrics: **a** input color hand image; **b** equivalent grayscale image; **c** actual binary hand contour with several smaller components; **d** largest and main contour component after rotation; **e** after wrist removal, final normalized hand; and **f** pictorial representation of computed features. The circle below the mid-finger is the centroid of the normalized hand

i.e., the local maxima, is considered as the tip of that finger. On the contrary, the radial distance map between palm contour point and a wrist reference point is considered for the same purpose [13]. Though, in terms of computation, the radial distance transform is time-consuming. Thus, the first approach is followed here. The outcomes of the key steps during preprocessing are shown in Fig. 2.

3.2 Geometric Feature Definition

Altogether fourteen features per finger are computed from every normalized hand image, some of which are pose-invariant in nature. The feature set includes

a. finger length: **1**
b. widths at three positions, one per phalanx: **3**
c. lengths of the major axis and minor axis of the ellipse fitted over the finger contour which are invariant to affine transformations: **2**
d. Euclidean distances from the centroid to eight equidistant contour points, which are invariant to translation: **8**

Large variation of the scalar values of the raw features implies to normalize features uniformly into a standard scale, generally in the range of [0, 1]. The z-score normalization is a popular method [23]. Here, the features are normalized to [0, 1] as it is required for multirange values of the features using the min–max approach. Otherwise, one feature with a higher-range value can dominate over other feature with a lower scalar value. Equation 3 is used for normalizing the raw features:

$$\bar{f}_{i,j} = \left(f_{i,j} - \min(f_i)\right) / (\max(f_i) - \min(f_i)) \tag{3}$$

where $f_{i,j}$ represents the ith raw feature of the jth subject, and $max(f_i)$ and $min(f_i)$ represent the maximum and minimum values of the ith feature, respectively. Now, the feature space (F) is defined by the normalized features $\left(\bar{f}_i\right)$ as $F = \{\bar{f}_i\}_{i=1}^{n}$, and the number of feature per finger is $n = 14$.

4 Feature Weighted *k*-NN Classification Algorithm

The k-nearest neighbor (k-NN) classifier is widely used in pattern classification due to its algorithmic simplicity [12, 18]. It determines the class membership of a test pattern based on a metric such as the Euclidean distance.

During the learning phase, the accuracy of each $\bar{f}_i$ is computed using a simple k-NN classifier which uses the Euclidean distance metric χ for determining the neighborhood. Based on the discriminative effectiveness, also known as relevance, a weight is assigned to every feature. The weight w_i of a feature $\bar{f}_i$ is assigned as $w_i = \frac{\chi(\bar{f}_i)}{\sum_{\bar{f}_i \in F} \chi(\bar{f}_i)}$, where $\chi(\bar{f}_i)$ represents the relevance of $\bar{f}_i$ evaluated by k-NN with respect to distance metric $\chi = \sqrt{\sum_{i=1}^{n} (T_{\bar{f}_i} - C_{\bar{f}_i})^2}$ and $\sum_{i=1}^{n} w_i = 1$. The trained and claimed test features are denoted by $T_{\bar{f}_i}$ and $C_{\bar{f}_i}$, respectively. Based on the weight assignment, the feature weighted k-NN (*fwk*-NN) classifier is excogitated. In the *fwk*-NN classifier, weighted distance metric is defined with the w_i values as

$$\chi' = \chi \cdot w_i = \sqrt{\sum_{i=1}^{m} (T_{\bar{f}_i} - C_{\bar{f}_i})^2 \cdot w_i} \tag{4}$$

Feature Weighted k-Nearest Neighbor (fwk-NN) Classification Algorithm

Input: trained featureset $T = \{T_{\bar{f}_i}\}_{i=1}^{n}$, claimed test feature set $C = \{C_{\bar{f}_i}\}_{i=1}^{n}$, class label $L = \{L_i\}_{i=1}^{p}$, threshold th_w

Output: recognized class label S

1. Calculate the relevance R of each feature independently by simple *k*-NN with distance metric χ

   $$\forall i \in n: \quad R(i) \leftarrow \chi(\bar{f}_i) \text{ where } \chi(\bar{f}_i) = \sqrt{\sum_{i=1}^{n} (T_{\bar{f}_i} - C_{\bar{f}_i})^2}, \ \bar{f}_i \in F, \text{and } 1 \le i \le n$$

2. Calculate the sum of all relevances $R = \sum_{i=1}^{n} R(i) = \sum_{i=1}^{n} \chi(\bar{f}_i)$ and assign a weight to every feature $w_i = \dfrac{\chi(\bar{f}_i)}{\sum_{\bar{f}_i \in F} \chi(\bar{f}_i)}$
3. Select $\bar{f}_i$ in subset F if $\forall i \in n: \ w_i \ge th_w$, discard otherwise.
4. Say, m features are selected in step 3. Define the weighted distance metric $\chi' = \chi \cdot w_i = \sqrt{\sum_{i=1}^{m} (T_{\bar{f}_i} - C_{\bar{f}_i})^2} \cdot w_i$
5. Calculate the distances χ' for all the L classes and selects the class with the minimum distance among the nearest n neighbors S=argmin$\{\chi'_{L_1}, \ldots, \chi'_{L_p}\}$
6. Return S

Now, the performances are evaluated with the *fwk*-NN classifier that uses metric χ'. A threshold value (th_w) is considered to select a feature independently whether $\bar{f}_i$ should be included in F or not, i.e., $\bar{f}_i \in F$ if $w_i \ge th_w$, otherwise $\bar{f}_i$ is rejected. Thresholding is used to eliminate insignificant feature which is defined as $\bar{f}_i$ that provides very poor accuracy while tested alone independent of other feature. Finally, the subset of relevant features is used for classifying the unknown test data.

5 Performance Evaluation

A framework has been developed for conducting experiments on the right- and left-hand 2-D images of 500 subjects of the Bosphorus database using the *fwk*-NN classifier. Detailed description of the database is available in [7].

In this dataset, only three hand images per subject are available. Thus, two feature vectors are trained and one vector is tested at a time. Three different combinations of training and testing vectors are generated by considering two different vectors for training, and remaining vector is for testing simultaneously. The average results of those unique combinations are reported below.

5.1 Experimental Description

Experiments are performed for the identification and verification modes. In the identification mode, the feature vector of a person is compared against the templates stored in the database. To establish user identity, a respective feature template is matched (one-to-many) with all the templates. A genuine user is accepted when the matching score is the maximum, and difference between enrolled and test feature set is lowest. In the case of an unknown identity, claimant features are tested for either false acceptance or true rejection. Based on the distance threshold (Eq. 5), it can be determined whether the identity is genuine or imposter. If the matching score lies within the threshold, then the claimer is accepted as legitimate, otherwise rejected as an imposter. The threshold is defined as

$$th_1 = \sqrt{\sum_{\bar{f}=1}^{n} \frac{(T_{\bar{f}} - C_{\bar{f}})^2}{(\sigma_{\bar{f}}^2)}} \tag{5}$$

where $\bar{f}_i$ represents every single feature, T denotes the trained feature of a subject, C represents any unknown identity of the claimer, and σ is the standard deviation of $\bar{f}_i$. Similarly, the weight of every feature is considered to determine the threshold. In this case, the weights used are same as calculated for the identification, given in Eq. 6:

$$th_2 = \sqrt{\sum_{\bar{f}=1}^{m} \frac{(T_{\bar{f}} - C_{\bar{f}})^2}{(\sigma_{\bar{f}}^2)} \cdot w(\bar{f})} \tag{6}$$

The verification performance is calculated in terms of the following:

False Accept Rate (FAR): An imposter matching score is lower than the threshold.
False Reject Rate (FRR): A genuine matching score is more than the threshold.
Genuine Accept Rate (GAR): 1-FRR.
Equal Error Rate (EER): It is the value at which the FAR equals to the FRR.

The GAR and the FAR depend on the decision threshold and are plotted in a Receiver Operating Characteristic (ROC) curve.

5.2 Result Analysis

The first experiment has been conducted using a simple k-NN classifier with $k=3$. The subjects are divided into subpopulations with different sizes of the random combinations of the left- and right-hand subjects. Several different combinations are formulated and tested. The average accuracies are reported in Table 4. In the next experiment, the accuracy of each is feature is computed in the learning phase using the k-NN based on which a w_i is assigned to every feature accordingly. Now, a threshold value th_w is considered to select only the good features satisfying $w_i \geq th_w$. The subset F of significant features is considered for the testing on the unknown feature vectors using the *fwk*-NN classifier. Three different subsets are tested which are determined based on th_w. It is obvious that a higher th_w selects a lesser number of features than a th_w with a lesser value. The accuracies (given in Table 5) imply that the *fwk*-NN performs better than the k-NN on this present dataset.

In the verification mode, two different tests are conducted. The EERs are computed and provided in Table 6. The first test considers th_1 (Eq. 5) for calculating the EER. In the second test, th_2 (Eq. 6) is used to compute the EERs. The results reflect that the weighted threshold th_2 performs better than simple threshold th_1. Finally, the ROC curves for the 500 subjects are plotted in Fig. 3.

The performances of the present work are comparable directly to other approaches experimented over the same dataset, and the population size is also same, presented in Table 7. The present geometric feature set is competitive over the shape-based features, mentioned in Table 7.

Table 4 Identification performance (%) using k-NN

Hand	Subjects				
	100	200	300	400	500
Right	100	98.5	97.34	96.25	96
Left	100	99.5	98.67	97.75	96.8

Table 5 Identification performance (%) using *fwk*-NN

Hand	Feature subset size	Threshold (th_w)	Subjects				
			100	200	300	400	500
Right	17	1.5	100	98.5	96.34	96	94.6
	41	1	100	99	97	96.75	96.6
	48	0.5	100	99.5	98.34	97.75	97
Left	28	1.5	100	98	97.34	95	94.6
	43	1	100	98.5	98	97	96.2
	46	0.5	100	99	98.34	97.25	97

Table 6 Verification performance: EER (%)

Hand	EERs of 500 Subjects	
	k-NN	*fwk*-NN
Right	6.31	5.94
Left	6.24	6.08

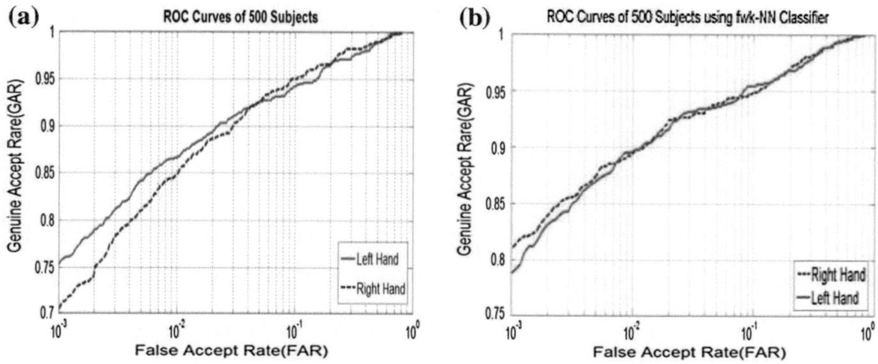

Fig. 3 ROC curves of the 500 subjects using **a** th_1 and **b** th_2 plotted in semilogarithmic scale for X-axis

Table 7 Performance comparison

Author	Method (hand)	Subjects	Accuracy (%)
Dutağaci et al. [7]	PCA on binary hands (left)	400	96.1
El-Sallam et al. [22]	Fusion-based ICA (right)	500	96.2
Proposed	Geometric (left)	400	97.75
	Geometric (both)	500	97

6 Conclusion

In this paper, recent state-of-the-art in hand biometrics has been reported with a focus on the key challenges that require research attention. A method for human identification using pose-invariant hand geometry is presented with the acceptable results, experimented over 500 individuals. The *fwk*-NN classifier performs better than simple *k*-NN classifier. Remarkable progress has been accomplished, and the challenges remain open for further advancement. Still, the challenges have to be dealt with the adaptation and human acceptability of the hand biometrics for a larger population in higher security applications to uplift at the next level of achievement. It can be expected that in near future this trait may find its suitability for high-security commercial mobile application and forensic investigation.

Acknowledgements The authors would like to thank Prof. B. Sankur of Bogazici University for providing the hand image database used in this paper.

References

1. Jain, A.K., Ross, A., Prabhakar, S.: An introduction to biometric recognition. IEEE Trans. Circuits Syst. Video Technol. **14**(1), 4–20 (2004)
2. Borgia, E.: The internet of things vision: key features, applications, and open issues. Comput. Commun. **54**, 1–31 (2014)
3. Jain, A.K., Nandakumar, K., Ross, A.: 50 years of biometric research: accomplishments, challenges, and opportunities. Pattern Recogn. Lett. **79**, 80–105 (2016)
4. Miller, R.P.: Finger dimension comparison identification system. U.S. Patent No. 3576538 (1971)
5. Beenau, B.W., Bonalle, D.S., Gray, W.J., Larkin, C., Montgomery, J.L., Saunders, P.D.: Hand Geometry Recognition Biometrics on a Fob. U.S. Patent No. 7886157 B2 (2011)
6. Kanhangad, V., Kumar, A., Zhang, D.: A unified framework for contactless hand verification. IEEE Trans. Informat. Forensics Secur. **6**(3), 1014–1027 (2011)
7. Dutağaci, H., Sankur, B., Yörük, E.: A comparative analysis of global hand appearance-based person recognition. J. Electron. Imag. **17**(1), 011018/1–19 (2008)
8. Duta, N.: A survey of biometric technology based on hand shape. Pattern Recogn. **42**(11), 2797–2806 (2009)
9. Kang, W., Wu, Q.: Pose-invariant hand shape recognition based on finger geometry. IEEE Trans. Syst. Man Cybernet. Syst. **44**(11), 1510–1521 (2014)
10. Sharma, S., Dubey, S.R., Singh, S.K., Saxena, R., Singh R.K.: Identity verification using shape and geometry of human hands. Expert Syst. Appl. **42**(2), 821–832 (2015)
11. Charfi, N., Trichili, H., Alimi, A.M., Solaiman, B.: Novel hand biometric system using invariant descriptors. In: IEEE International Conference on Soft Computing and Pattern Recognition, pp. 261–266 (2014)
12. Vivencio, D.P., Hruschka, Jr. E.R., Nicoletti, M.D., Santos, E.B., Galvão, S.D.C.O.: Feature-weighted k-Nearest Neighbor Classifier. IEEE Symposium on Foundations of Computational Intelligence (FOCI), pp. 481–486 (2007)
13. Yörük, E., Konukoğlu, E., Sankur, B., Darbon, J.: Shape-based hand recognition. IEEE Trans. Image Process. **15**(7), 1803–1815 (2006)
14. Wang, M.H., Chung, Y.K.: Applications of thermal image and extension theory to biometric personal recognition. Expert Syst. Appl. **39**(8), 7132–7137 (2012)
15. Faundez-Zanuy, M., Mekyska, J., Font-Aragonès, X.: A new hand image database simultaneously acquired in visible, near-infrared, and thermal spectrums. Cogn. Comput. **6**(2), 230–240 (2014)
16. Ferrer, M.A., Morales, A., Diaz, A.: An approach to SWIR hyperspectral hand biometrics. Informat. Sci. **268**, 3–19 (2014)
17. Morales, A., Ferrer, M.A., Cappelli, R., Maltoni, D., Fierrez, J., Ortega-Garcia, J.: Synthesis of large-scale hand-shape databases for biometric applications. Pattern Recogn. Lett. **68**(1), 183–189 (2015)
18. Luque-Baena, R.M., Elizondo, D., López-Rubio, E., Palomo, E.J., Watson, T.: Assessment of geometric features for individual identification and verification in biometric hand systems. Expert Syst. Appl. **40**(9), 3580–3594 (2013)
19. Hu, R.X., Jia, W., Zhang, D., Gui, J., Song, L.T.: Hand shape recognition based on coherent distance shape contexts. Pattern Recogn. **45**(9), 3348–3359 (2012)
20. Guo, J.M., Hsia, C.H., Liu, Y.F., Yu, J.C., Chu, M.H., Le, T.N.: Contact-free hand geometry-based identification system. Expert Syst. Appl. **39**, 11728–11736 (2012)

21. Travieso, C.M., Ticay-Rivas, J.R., Briceño, J.C., Pozo-Baños, M., Alonso, J.B.: Hand shape identification on multirange images. Informat. Sci. **275**, 45–56 (2014)
22. El-Sallam, A., Sohel, F., Bennamoun, M.: Robust pose invariant shape-based hand recognition. In: 6th IEEE Conference on Industrial Electronics and Applications, pp. 281–286 (2011)
23. Hanmandlu, M., Grover, J., Madasu V.K., Vasirkala, S.: Score level fusion offhand based biometrics using T-norms. IEEE, 70–76 (2010)

Pre-bond Testing of TSVs in 3D IC Using Segmented Cellular Automata

Bidesh Chakraborty and Mamata Dalui

Abstract Pre-bond testing of through silicon vias (TSVs) in 3D ICs is very challenging task. Reliability and correctness of 3D ICs consisting of numbers of TSVs have to be established. This paper presents an innovative pre-bond test method for TSVs in a 3D IC. The logic is devised around the cascadable arrangement of Cellular Automata (CA), discovered by von Neumann in the year of 1960. SACA, a designated class of CA, has been developed to diagnose conductor and insulator defects in TSVs. The segmentation of CA in the design assures more effectiveness, with respect to the number of computations to identify faulty TSVs. The implemented hardware realizes quick test of manufacturing TSV faults in the pre-bond phase.

Keywords Pre-bond test · 3D IC · TSV · Fault detection · SACA

1 Introduction

3D ICs constructed with multiple dies arranged vertically and connected by through silicon vias (TSVs) are in place to offer more benefits over 2D ICs [1]. 3D ICs featuring minuscule form factor and reduced cost of manufacturing assures high-performance gain with the aid of reduced length of global interconnect. Some factors, e.g., thermal management, yield improvement, infrastructure construction, and so on for a 3D IC throws challenges over 2D ICs and eventually makes 3D IC testing more challenging.

Despite the advancements in 3D IC technologies, testing of TSVs incures a major concerns of researchers. Testing of TSVs can be made before bonding that is pre-bond phase and after bonding that is post-bond phase [2, 3]. Testing TSVs in pre-bond detects the defects arise during manufacture, such as conductor defects, insulator

B. Chakraborty (✉)
Haldia Institute of Technology, Haldia 721657, WB, India
e-mail: bidesh.me@gmail.com

M. Dalui
National Institute of Technology Durgapur, Durgapur 713209, WB, India
e-mail: mamata.06@gmail.com

P. Chandra et al. (eds.), *Information Technology and Applied Mathematics*,
Advances in Intelligent Systems and Computing 699,
https://doi.org/10.1007/978-981-10-7590-2_9

defects. On the other hand, post-bond testing identifies faults caused by different integration steps, e.g., thinning and bonding.

Chen et al. [4], presents a BIST architecture to test TSVs before bonding, using sense amplification technique. This technique consists of a test module (TM) associated with all TSVs, a Test controller that generates signals to be broadcast to each TM, a Test Result Collector (TRC) collects values from the flip-flops (FF) associated with each TM. Each TM consists of a tristate buffer, an NMOS transistor, and two cascade inverters as write driver, pull-down circuit, and the sense amplifiers, respectively. The testing circuit takes huge area and that is the overhead of the technique.

In this current work, we implement a test logic around CA theory to detect faulty TSVs before bonding. The unconventional solution is based on the single length cycle single attractor CA (SACA) [5–7] which is a designated class of CA [5] that leads to reduction in area of test hardware as well as total computation cost. The cascadable arrangement of CA makes the design highly scalable.

2 Fundamentals of Cellular Automata (CA)

A Cellular Automaton (CA) can be viewed as an autonomous finite-state machine (FSM). A CA cell is having two states—0 or 1 and the next state (NS) of ith CA cell is $Q_i^{t+1} = F_i(Q_{i-1}^t, Q_i^t, Q_{i+1}^t)$, where Q_{i-1}^t, Q_i^t, and Q_{i+1}^t are the present states (PS) of the left neighbor, self, and right neighbor of the ith cell at time t. The $Q^t = (Q_1^t, Q_2^t, \cdots, Q_n^t)$ at time t is the PS of CA.

The F_i is the next state function can be demonstrated in the form of a truth table (Table 1). A rule R_i [5] is the decimal equivalent of the eight outputs shown in Table 1. There are 2^8 (256) rules in a 2-state 3-neighborhood CA. Three such rules 240, 207, and 192 are reported in Table 1. Rule min term (RMT) at time t is the possible 2^3 (8) combinations of present states of (i−1)th, ith, and (i + 1)th cells, (1st row of Table 1). Column 101 of Table 1 is the 5th RMT. The NS corresponding to this RMT is '0' for rule 192 and 207 and '1' for rule 240. The last three rows (Table 1) show the NS of the ith cell at (t + 1), forming the rules 240 (NS = Q_{i-1}), 207 (NS = $Q'_{i-1} + Q_i$), and 192 (NS = $Q_{i-1}.Q_i$).

A set of rules R = $< R_1, R_2, \ldots, R_i, \ldots, R_n >$ that configure the CA cells is known as rule vector. If the rule vector contains all same rules, then the vector is

Table 1 RMTs of CA $< 240, 207, 192 >$

PresentState	111	110	101	100	011	010	001	000	*Rule*
RMT	7	6	5	4	3	2	1	0	
NextState	1	1	1	1	0	0	0	0	240
NextState	1	1	0	0	1	1	1	1	207
NextState	1	1	0	0	0	0	0	0	192

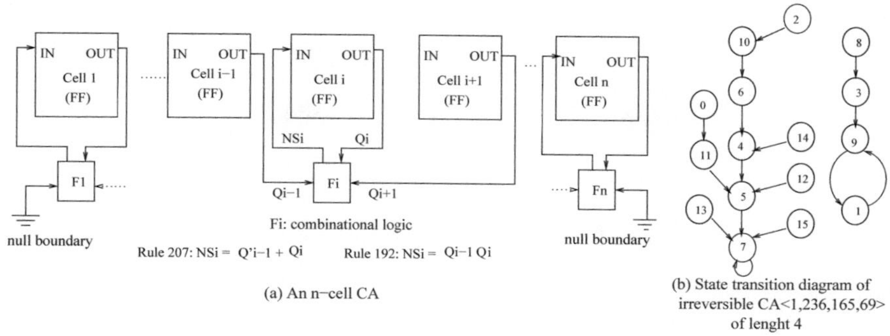

Fig. 1 Block diagram of null boundary n-cell CA

recognized as uniform CA; elseways, the vector can be said as hybrid/non-uniform CA. If the right neighbor of the rightmost and left neighbor of the leftmost terminal cell is permanently fixed to 0-state, then the CA is called null boundary CA. This class of CA is considered for our current work. Based on the state transition diagram of CA, we classify CA into two category, i.e., reversible CA and irreversible CA. If the states of a CA only form cycles in its state transition diagram, it is referred to as reversible; otherwise, the CA is irreversible (Fig. 1b). A set of states that CA forms cycle (loop) ($7 \rightarrow 7$ and $9 \rightarrow 1 \rightarrow 9$ of Fig. 1b) is defined to as the *attractor* or *point* state. The attractors with cycle length one, that is, $7 \rightarrow 7$ of Fig. 1b are of our major concern.

3 Defects in TSVs Before Bonding

At pre-bond phase, one end of the TSVs is linked to a metal layer, whereas the another end is floating and insulated from the substrate by a surrounding dielectric layer. Defects that can occur in the pre-bond phase are explained next.

Conductor Defects—This type of defects occurs if a TSV has a break in the manufacturing time. The break may or may not be open entirely, but it can act as a high resistive path (Fig. 2a). These prevent the signal to pass through the TSVs.

Insulator Defects—These defects are created by particles Induced by impurities or dust at the time of fabrication procedure. Such type of defect constructs a path between substrate and the TSV shown in Fig. 2b. This can be represented as a resistive short defects. This creates a bypass to substrate and causes the signal delay that leads to transition fault.

In both the defects, the signal does not transmit through TSV in specified time.

Induced Defects—These defects can be caused by the induction of the neighboring TSVs. This generates a signal in the receiving end of the TSV when there is actually no signal from the sender side of the TSV.

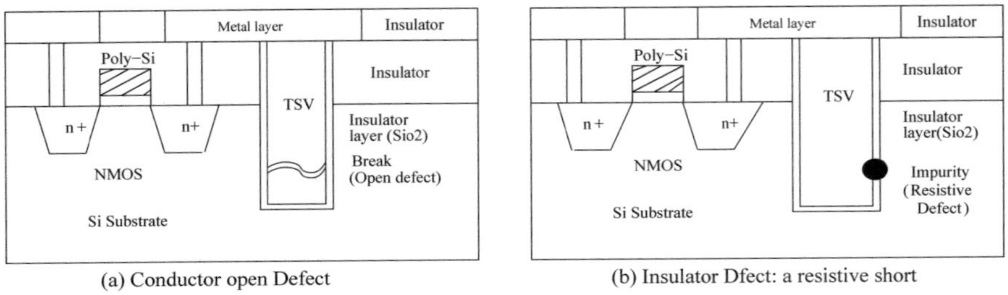

Fig. 2 Defects at manufacturing phase [4]

4 Proposed Design

The proposed model of pre-bond TSV test scheme designed around CA is shown in Fig. 3. Here, one end of the TSVs (n numbers in a row can be tested at a time) is linked to a metal layer, whereas the other end is on the surface, connected to the n-cell CA module. In case of both the conductor and insulator (CI) defects, the signal is prevented to pass through the TSVs in specified time. To test both the defects (CI), the TSVs connected with a die transmit all 1's to the other end of the TSV. Using the received pattern, the CA-based test unit associated to the other ends of the TSVs is configured with suitable rules and is then run for t (=n) time steps. The least significant bit (LSB) of the attractor identifies CI defects in single or multiple TSVs. The value of the LSB 1 leads to single or multiple faults in each TSV. Similarly for testing Induced defects, we transmit all 0's and the value of LSB 1 leads to single or multiple Induced faults.

Design Based on CA Cluster: The design reported so far needs t (= n, reported in Sect. 5.1) time steps to verify the TSVs. The cluster of CA makes the system more compact and achieves a quicker decision. For n TSVs, we use an (n+2^{p+1})-cell CA which is clustered into 2^p segments. (n + 4)-cell CA will be clustered into two segments (CA1 and CA2). Each of the CA segment (CA1 or CA2) consists of of $m = \frac{n}{2} + 2$ cells.

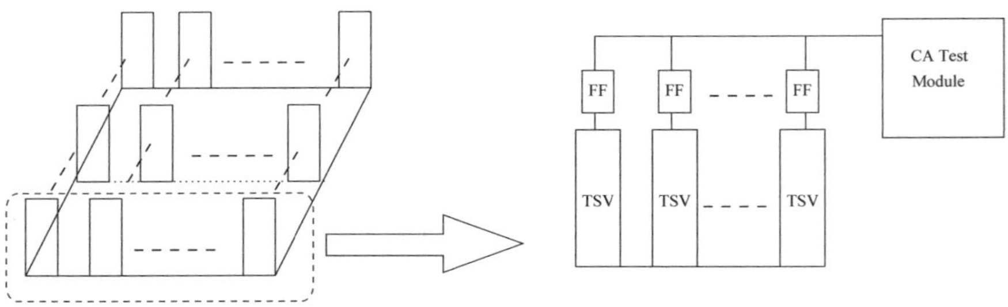

Fig. 3 Test Design of TSVs in Pre-bond phase

For a system with n number of TSVs and p = 1, the status information from C_1, C_2 ...$C_{n/2}$ constructs CA1 and $C_{n/2+1}$...C_n constructs CA2 (Fig. 5a). Both segments compute n/2 steps instead of n steps and taking OR operation of LSB1 and LSB2 says whether there is faults OR not.

5 The Formation of SACA

As the NS of SLCA is same as the present state, there must be one or more RMTs (Sect. 2) of a cell rule R_i of R for which the cell i retains its present state in the successive time steps. The cell i associated with the rule R_i retains its state if the RMTs 0, 1, 4, 5 are 0. On the other hand, if the RMTs 2, 3, 6, or 7 are 1 in a rule R_i, then the rule also retain to its current state in the successive time steps.

Theorem 1 *A CA rule R_i can be a probable to contribute to form SLCA(s) if at least one of its 2, 3, 6, or 7 is 1 and/or at least s 0, 1, 4, or 5 is 0 [7].*

All the 256 CA rules can be classified [7] in 9 groups (group 0–8) following Theorem 1. A rule has a higher probability to form SLCA(s), if most of the RMTs of the CA cell rule(s) follow Theorem 1.

It has been noticed that to construct an SACA, the rules must follow Theorem 1. Though, a rule 204 maintains Theorem 1 for all its RMTs cannot form an SACA.

Theorem 2 *If a CA rule 'r' denies Theorem 1 for some RMTS, it can construct an uniform SACA with rule 'r', the 'r' must deny Theorem 1 for some RMTs [7].*

Rule 48 belonging to group 2 denies Theorem 1 for 6 RMTs Table 2 of [7]. Further, rule 192 belonging to group 6 denies Theorem 1 only for 2 RMTs Table 2 of [7]. Each of these rules forms SACA for all lengths.

Theorem 3 *A uniform SACA rule R_o when hybridized by a rule R_h, it can produce a new attractors iff the intersection between the set of RMTs of R_o and R_h, for which Theorem 1 is denied is null.*

For example, RMTs {2, 3} and RMTs {0, 1} of the rules 192 and 207, respectively, for which Theorem 1 denies can follow Theorem 3. That is, the CA generated from rule 192 (uniform SACA) due to defects at single or multiple nodes hybridized by rule 207 (hybrid CA). The combinations produce multiple attractors and when initialized with all 1 s seed, fixes to an point or attractor state with LSB = 1 (Fig. 4b, c).

5.1 *Test Design Using SACA*

To test conductor and insulator defects, the TSVs (TSVs organized in a row of a die) send all 1's and to test Induced defects it transmit all 0's to the flip-flops associated with each TSVs. Using the accumulated bits in the receiving end, the CA-based test

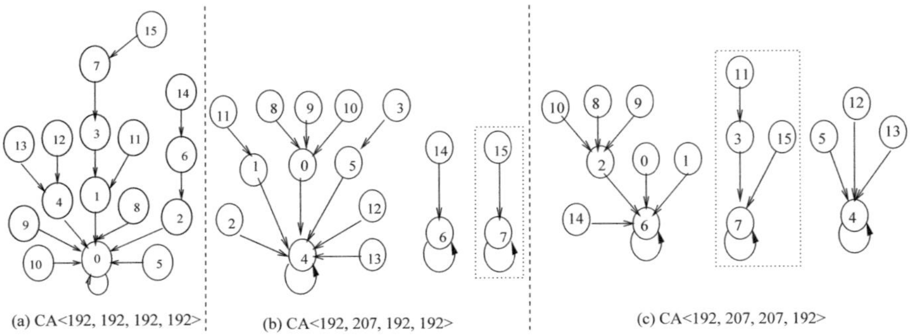

Fig. 4 State transition diagrams of uniform and hybrid CA

unit attached to the receiving end of the TSVs is set with appropriate rule and is execute for n number of times where n is equal to the CA cell numbers. The CA fixes to an attractor/point state and the LSB of the attractor (state of least significant CA cell) indicates corresponding defects in single or multiple TSVs.

(1) Conductor and Insulator Defects (CI): For detection of CI faults, the test vector consisting of a sequence of all 1's ('11...1') will be send through the TSVs. The received bits stored in FFs associated with each TSV are directly utilized to fix the rule of CA cell. When all the TSVs are fault-free, then the FFs perfectly get the transmitted bit sequence ('11...1') and the CA devised is a uniform SACA set with the rules $< 192, 192, .., 192 >$. On the other hand, when one or more TSVs are affected with CI faults, the respective received bits within the received sequence are flipped to 0. Accordingly, the CA formed is a hybrid one hybridized with rule 207 at the corresponding CA cell(s). As the CA rules are decided, it is then run for n-time steps and finally fixes to an attractor state with LSB 0 (Fig. 4a) in fault-free case and LSB 1 (Fig. 4b, c) for a faulty case pointing out the existence of faulty TSVs.

Theorem 4 *A non-uniform CA configured with rule 192 and 207, when run with an all 1 s initial seed fixes to an attractor/point state with LSB 1.*

Figure 4b, c shows the diagrams of two different CA sequence (base rule is 192) hybridized with rule 207. Both the cases, the CA fixes to an attractor with LSB 1 when initiated with all 1 s seed and runs for n (=4 because the number of CA cells are 4)-time steps. For example, in Fig. 4b it fixes to attractor/point state 0111 (state '7').

(2) Induced Defects: For detection of Induced fault, the test vector consisting of a sequence of all 0's ('00...0') will be send through the TSVs. The received bits stored in FFs associated with each TSV are directly utilized to fix the rule of CA cell. If a FF stores 0(1), it fixes the rule 192 (207). When all the TSVs are fault-free, then the FFs perfectly get the transmitted bit sequence ('00...0') and the CA devised is a uniform SACA fix with the rules $< 192, 192, .., 192 >$. On the other hand, when one or more TSVs are affected with Induced faults, the respective received bits within the received sequence are flipped to 1. Accordingly, the CA formed is a hybrid one hybridized with rule 207 at the corresponding CA cell(s). As the CA rules are decided, it is then run for n-time steps and finally fixes to an attractor/point state with LSB 0 (Fig. 4a) in fault-free case and LSB 1 (Fig. 4b, c) for a faulty case pointing out the existence of faulty TSVs.

6 Hardware Realization of the Scheme

Figure 5b depicts the hardware realization of the test scheme. Here, CA cell's next state is generated by a combinational logic, consists of a 4-to-1 multiplexer (MUX) and few basic gates. The ith CA cell rule is selected by the MUX. The MUX inputs are 0(1) to detect CI (Induced) faults and received bit (ith). The MUX input 01 considers the ith bit is 1 for CI fault, and the rule enabled for the cell (ith) is 207 ($Q'_{i-1} + Q_i$). Similarly when the ith bit (received) is 0 to test same fault (CI), then the MUX input will be 00 and the rule enabled for ith cell will be 192 ($Q_{i-1}.Q_i$). Further for Induced faults, if the ith bit is 0(1), i.e., the MUX input is 10(11), then the rule enabled for ith cell is 207(192), just opposite rules of CI fault testing. After n number of computations, the value of the LSB (1(0) for faulty (non-faulty)) leads to single or multiple CI or Induced faults in each TSV rows.

We can reduce the number of computations using segmented CA. For example, n number of TSVs can be divided into two parts. Two CA segment of length n/2 can used to run parallely to reduce number of computation (half) (Fig. 5a). After the computations, OR operation between the LSBs can say whether the rows of TSV have an error or not.

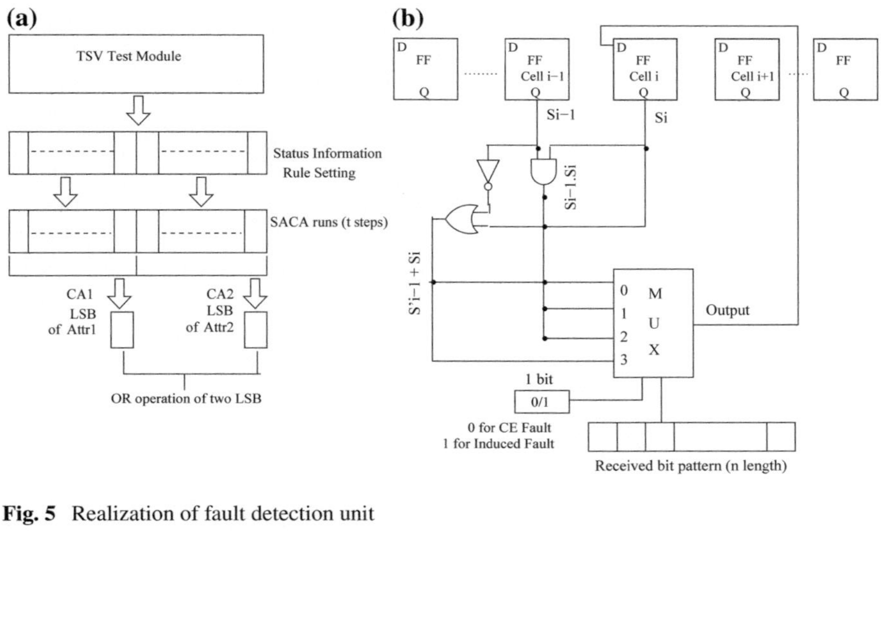

Fig. 5 Realization of fault detection unit

7 Result Analysis

In this section, we demonstrate evaluated results that establish the efficiency of our proposed solution for identifying defective TSVs in pre-bond phase. In our experimentation, we consider a die with 5–512 TSVs row wise (column 1 of Tables 2 and 3). Column 2 in Tables 2 and 3 reports the data that are transmited through the TSVs. Column 4 of Tables 2 and 3 depicts the rules configured based on the value stored in FFs associated with each TSVs in a row. To test CI faults, if the value of a FF is 1 then it is configured with rule 192; otherwise, the cell is configured with rule 207. For Induced faults, the rule configuration is just the opposite, i.e., 192 for 0 and 207 for 1. For testing of CI faults, the TSVs transmit all 1's as reported in column 2 of Table 2. Column 3 reports the received data through TSVs. Column 4 reports the rules that are configured. The attractor state (final state) and the LSB of attractor depicted in columns 5 and 6, respectively. The faulty/non-faulty status of TSVs is decided by checking the LSB of attractor. If the LSB of attractor is 0, then there is no CI fault in that row of TSVs; otherwise, the row of TSVs is affected either by conductor open or insulator defects (CI faults). Similarly, from Table 3, it can be sensed that if the LSB of the attractor is 1 then Induced fault has occurred in the row of TSVs; otherwise, no such fault is present.

Table 2 Evaluated results for conductor and insulator defects

TSV numbers	Data sent	Data received	Rule configuration	Attractor/point state	LSB of attractor	TSV status
5	11111	11001	192, 192, 207, 207, 192	00111	1	Faulty
5	11111	10101	192, 207, 192, 207, 192	01111	1	Faulty
5	11111	11111	192, 192, 192, 192, 192	00000	0	Non-faulty
10	11...11	11…11	192,192,…,192	00...00	0	Non-faulty
10	11...11	10…10	192, 207,..., 192, 207	01…01	1	Faulty
512	1111…1111	1001…0101	192, 207, 207, 192, ...207, 192, 207, 192	1	1	Faulty

Table 3 Evaluated results for Induced defects

TSV numbers	Data sent	Data received	Rule configuration	Attractor/point state	LSB of attractor	TSV status
5	00000	00110	192, 192, 207, 207, 192	00111	1	Faulty
5	00000	01010	192, 207, 192, 207, 192	01111	1	Faulty
5	00000	00000	192, 192, 192, 192, 192	00000	0	Non-faulty
10	00…00	00…00	192, 192,…,192	00…00	0	Non-faulty
10	00…00	01…01	192, 207,…,192, 207	01…01	1	Faulty
512	0000…0000	0110…1010	192, 207, 207, 192,…207, 192, 207, 192	…1	1	Faulty

8 Conclusion

In this current work, we propose an effective test for detection of faulty TSVs before bonding. With this scheme conductor, insulator and Induced defects of TSVs can be tested with minimum computational and hardware overhead assuring robust and scalable design. The test block takes small amount of spaces, which is very significant in 3D testing. The design is made around the cascadable and modular structure of CA (Cellular Automata). The introduction of segmented CA approach ensures more effectiveness in the design, with resepect to computation counts. This ensures simpler hardware realization that leads to rapid identification of faulty TSVs before bonding.

References

1. Pavlidis, F., Friedman, E.G.: Interconnect-based design methodologies for three-dimensional integrated circuits. Proc. IEEE **97**(I), 123–140 (2009)
2. Mariniseen, E.J., Zorian, Y.: Testing 3D chips containing through silicon vias . In: International Test Conference, vol. E1.1 (2009)
3. Chakraborty, B., Dalui, M.: A cellular automata based test scheme for TSVs in 3D ICs. In: ICDCS, pp. 723–727 (2012)
4. Chen, P.-Y., Cheng, W., Kwai, D.-M.: On-chip TSV testing for 3D IC before bonding using sense amplification. In: Proceedings of IEEE Asian Test Symposium, pp. 450–455 (2009)
5. Wolfram, S.: Cellular Automata and Complexity—Collected Papers. Addison Wesley (1994)
6. Chaudhuri, P.P., Roy, D., Chowdhury, S., Nandi, S., Chatterjee, S.: Additive Cellular Automata–Theory and Applications, vol. 1. IEEE Computer Society Press, California, USA. ISBN 0-8186-7717-1 (1997)
7. Sukanta, D., Nazma, N.N., Sukanya, M., Mamata, D., Biplab, K.S.: Characterization of CA Rules For SACA Targeting Detection of Faulty Nodes in WSN. ACRI (2010)
8. Marinissen, E.J., Verbree, J., Konijnenburg, M.: A structured and scalable test access architecture for TSV-based 3D stacked ICs. In: Proceedings of Test Symposium, pp. 269–274, May 2010

Natural Language Description of Surveillance Events

Sk. Arif Ahmed, Debi Prosad Dogra, Samarjit Kar and Partha Pratim Roy

Abstract This paper presents a novel method to represent hours of surveillance video in a pattern-based text log. We present a tag and template-based technique that automatically generates natural language descriptions of surveillance events. We combine the output of some of the existing object tracker, deep learning guided object and action classifiers, and graph-based scene knowledge to assign hierarchical tags and generate natural language description of surveillance events. Unlike some state-of-the-art image and short video descriptor methods, our approach can describe videos, specifically surveillance videos by combining frame-level, temporal-level, and behavior-level target tags/features. We evaluate our method against two baseline video descriptors, and our analysis suggests that supervised scene knowledge and template can improve video descriptions, specially in surveillance videos.

Keywords Video to text · Surveillance video description · Video description

1 Introduction

Describing videos using natural language is an interesting domain of research. It can be used as text-based video searching, indexing, scene descriptor for blind, and an event descriptor in visual surveillance. It can present hours of videos using

---

Sk. Arif Ahmed (✉) · S. Kar
National Institute of Technology Durgapur, Durgapur 713209, WB, India
e-mail: arif.1984.in@ieee.org

S. Kar
e-mail: samarjit.kar@maths.nitdgp.ac.in

D. P. Dogra
Indian Institute of Technology Bhubaneswar, Bhubaneswar 751013, Odisha, India
e-mail: dpdogra@iitbbs.ac.in

P. P. Roy
Indian Institute of Technology Roorkee, Roorkee 247667, Uttarakhand, India
e-mail: proy.fcs@iitr.ac.in

P. Chandra et al. (eds.), *Information Technology and Applied Mathematics*,
Advances in Intelligent Systems and Computing 699,
https://doi.org/10.1007/978-981-10-7590-2_10

short textual descriptions. Representing images with a single sentence has received notable attention of the researchers [2, 5, 10, 11, 13] to generate natural language from video sequences [6, 16, 17, 20, 24]. The system can be beneficial for creating automated video descriptions or ground truth labeling. In this work, we propose a new approach to generate story-based short descriptions of moving targets appearing in a surveillance environment. Our approach performs surveillance video description using a hierarchical tag and template-based "if-then" model introduced by human experts. It incorporates a supervised scene model and generates a sequence of words in arbitrary length to describe surveillance events.

To the best of our knowledge, this is probably the first approach to surveillance video description log generator that uses hierarchical deep learning framework to generate textual descriptions of surveillance events. It involves a video-to-text generation method based on predefined tags and templates. First, moving objects are detected, classified, and assigned object-specific tags such as type, size, and color. All tags are an output of a convolutional neural network (CNN). Next, objects are tracked and the scene-specific tags such as speed and movement pattern are extracted. Finally, the set of tags are arranged based on template to describe surveillance videos. For instance, a typical target in our dataset contains tags {white,vehicle,medium,normal speed,entry from main building,exit to residential zone} generates natural language description based on template "A {color} {size} {type} in {speed}, coming from {entry zone} toward {exit zone}". The output description is generated as "A white medium vehicle in normal speed, coming from Main Building toward Residential Zone" as depicted in Fig. 1.

Rest of the paper is organized as follows. Section 2 discusses the existing work in this domain of research. In Sect. 3, we describe the proposed tag-based

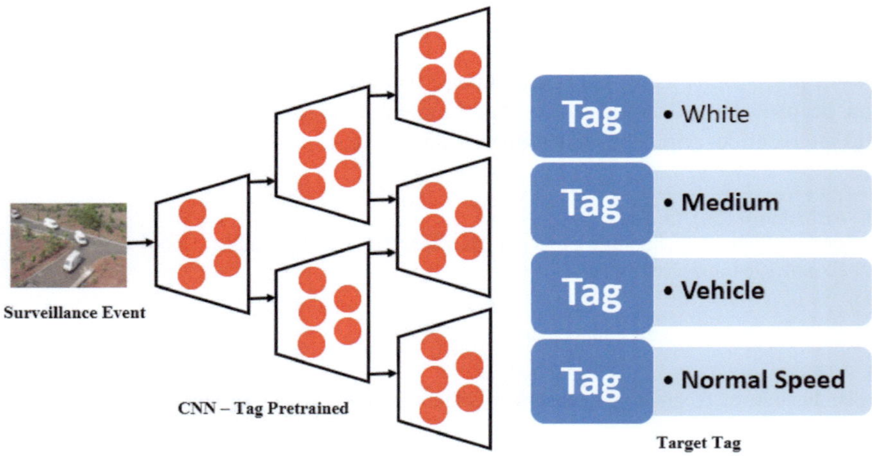

Fig. 1 Target tag: white, medium, vehicle, entry: main building, exit:residential zone, speed:normal. Text Log: *09:05 AM : A white medium vehicle in normal speed, coming from Main Building toward Residential Zone*

description generation process. Section 4 presents results using public as well as in-house datasets. Finally, Sect. 5 concludes the paper.

2 Related Work and Contributions

Early work on video-to-natural language such as metadata-based [1], caption-based sentence generator [6, 12] uses vision and text mining approach. The problem of generating textual descriptions of surveillance videos is difficult due to the presence of varying sets of objects, different patterns of movements, scenes, attributes, and their behaviors. Early work on video captioning by extracting metadata and clustering similar captions [9, 22] has been used in query-based video retrieval. Existing approaches for generating natural language descriptions [6, 12] have been used to identify the semantic content, such as subject, object, and verb, and then generates a sentence based on a predefined rule. In this field of research, long short-term memory (LSTM) networks [19] have successfully applied in sequence analysis. LSTM is a recurrent neural network (RNN) that enables learning of the language model. Most of the existing literature primarily focuses to generate action-specific description. However, due to large amount of action, performance of these methods may not be satisfactory (less than 30%) always. They do not incorporate scene-specific knowledge such as region information or long-duration behavior like loitering.

Our main contributions are as follows: (a) Shifting from the RNN-based learning such as LSTM, we have proposed a hierarchical neural network to generate textual descriptions of surveillance videos, and (b) unlike frame-level- and temporal structure-based text generation methods, we have used frame-level, temporal-level, and scene level information to generate natural language text.

To archive this, we assume (a) the surveillance scene background is represented by a set of known regions, and (b) targets are allowed to move region to region and can perform a set of activities.

3 Proposed Tag-Based Description Generation

Our overall method for generating textual log is depicted in Fig. 2. We assume a surveillance scene with some prior region information. We have used three types of information to generate the descriptions, namely frame-level information such as target type, short-term behavior information, and long-duration scene activity like activity regions. First, moving objects are tracked. Next, the targets are classified into different classes, such as pedestrian, car. The behavior of the targets is represented using a graph-based pattern discovery [4]. Finally, a sentence is generated by direct mapping to the template provided by human experts.

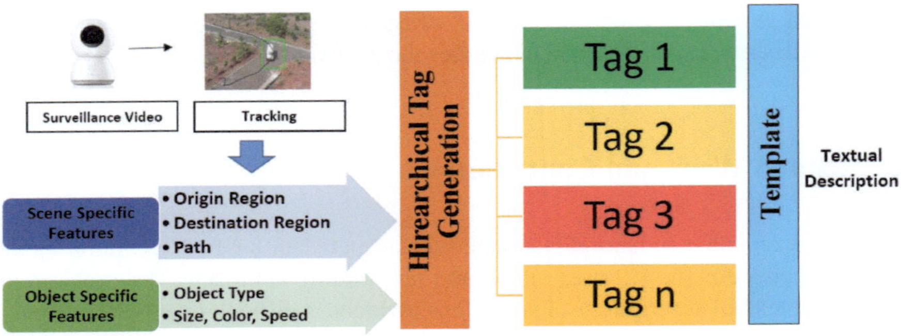

Fig. 2 Flow of the proposed description/ground truth generation method. First, moving objects are detected, classified, tracked, and assigned tags from a known set of tags. Next, the tags are arranged as per a predefined template to generate descriptions of the events

Fig. 3 Four regions marked by experts in our dataset. Regions are R1: Main Building, R2: Middle Road, R3: Main Gate, and R4: Residential Zone

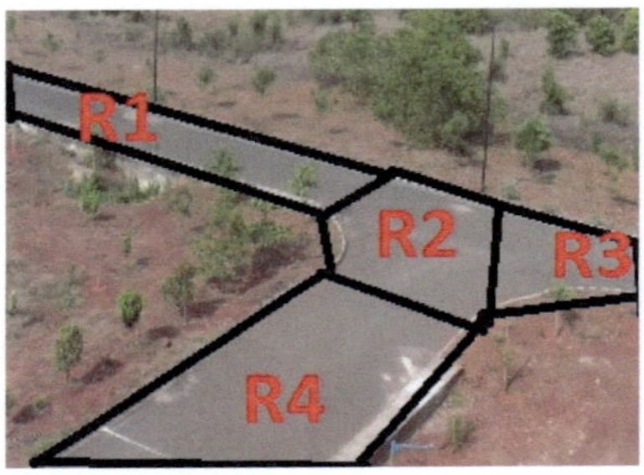

3.1 Surveillance Video and Text Representation

Scene Knowledge: A given surveillance scene is defined by a set of regions. A moving target is supposed to move from one region to another. Unspecified regions are marked as forbidden zones. For example, Fig. 3 presents the regions of our dataset. The details of the dataset are discussed in result section. There are four regions ($\{R1, R2, R3, R4\}$) specified by the expert. The regions are named as {Main Building, Middle Road, Main Gate, Residential Zone}.

Target Tracking: Our proposed system is based on object-specific features such as speed, size, long-duration activity. Hence, segmenting background/foreground and long-duration tracking is one of the important prerequisites for generating textual descriptions. Inaccurate object tracking may lead to poor quality descriptions. During real-time, moving objects are detected using improved mixture of Gaussian [25] and then tracked using Kalman filter-based multi-object tracker [23]. Initially, moving objects are detected as described earlier and represented by bounding box (x_t, y_t, w_t, h_t). Each detection is associate with a target (τ). τ as given in (1),

$$\tau = [u, v, s, a, \dot{u}, \dot{v}, \dot{a}]^T \tag{1}$$

Fig. 4 A target (vehicle) in IIT dataset and segmentation results at different frame

where u, v, s, and a represent the horizontal and vertical motions of the targets area and aspect ratio of the targets bounding box, respectively. $\dot{u}$, $\dot{v}$, $\dot{s}$, and $\dot{a}$ are derivative with respect to time. Hungarian optimization algorithm has been used to assign each detection with a target. If the target is lost for a certain number of frames, kernelized correlation filter (KCF) [7] has been used to search the lost object and update target state. KCF tracker is a single image trained classifier. A trajectory is defined by (2), where (x_i, y_i) is the position of the target, c_i is the segmented contour in i'th frame, and n is the length of the trajectory.

$$\tau = \{< x_1, y_1, c_1, t_1 >, < x_2, y_2, c_2, t_2 >, ..., < x_n, y_n, c_n, t_n >\} \tag{2}$$

Figure 4 shows positions and segmentation of a moving target in IIT dataset video.

Behavior Understanding: Behavior of a target is defined by its region-to-region movement. First, the trajectory (τ) is represented by a sequence of known regions in the surveillance environment, for instant $R1$, $R1$, $R1$, $R2$, $R2$, $R3$, $R3$. Next, the target is represented by entry and exit regions using (3), e.g., ($R1$, $R3$).

$$\tau = \{R_{entry}, R_{exit}\} \tag{3}$$

Deep Learning Guided Object Tagging: Unlike existing LSTM-based image captioning approaches [5, 21] and video-to-text generator [20, 24], we apply a hierarchical convolutional neural network (CNN) to the input images and provide the output as a tag generator. In this work, we report results using three sub-layers as convolutional, max-pooling, and ReLU activation layer. A convolutional layer consists of a set of parameters such as mapping factor, kernel sizes, output size. In our designs, the network comprises of three convolutional layers and two fully connected layers. The first convolutional layer filters $32 \times 32 \times 3$ RGB input image for object classification with 32 kernels of size $5 \times 5 \times 96$. In the next layer, a normalized and pooled ($16 \times 16 \times 32$) output is filtered with 32 kernels of size $5 \times 5 \times 1024$. Third convolution is connected with $5 \times 5 \times 2048$ and pooled at ($8 \times 8 \times 32$). The fully

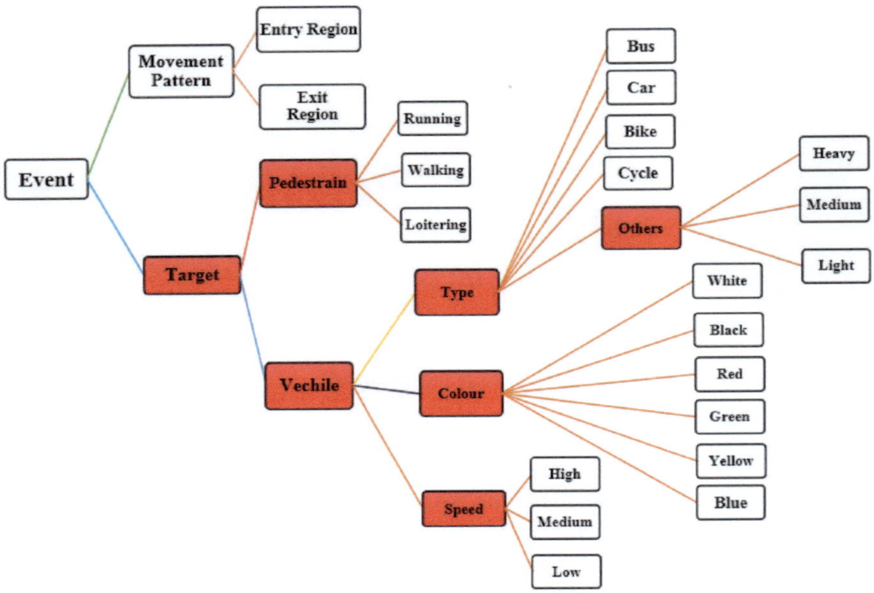

Fig. 5 Hierarchical classification of target behaviors and appearances. Here, each node represents a tag and red node represents CNN. Attributes include entry-to-exit patterns of movement, pedestrian behavior, vehicle type, size, speed, and color. These labels/tags are used to classify different characteristics of surveillance events, and we have used them in the procedural framework. Note that the list above is not exhaustive, as we can further extend attributes like interaction and action

connected layers have ($4 \times 4 \times 32$) inputs and 1024 and 64 number of neurons. A binary mask is used for size estimation. We have used CNNs that are pretrained using a 5 h in-house dataset video. The CNN is based on tiny-dnn, a small and faster CNN for classifications which is available freely in.[1] Moving targets are specified by a set of known objects, such as {Pedestrian, Car, Bus, Truck, Bike, Cycle, Vehicle} and known size {heavy, medium, light}. Figure 5 depicts the tag hierarchy used in our experiments, where each red node represents a CNN. Each combination of tags is directly mapped to existing templates. Figure 5 shows the hierarchical tag representation of our proposed method.

Generation of Natural Texts: We have used a template-based method to generate texts to represent targets and the behavior. First, a target is tracked and object-specific features such as target type, speed, and color are extracted hierarchically. Next, the region-to-region movement pattern is extracted by the trajectory of the object, and we assign entry-to-exit action tags such as "R1-to-R2". We define a set of all possible textual templates based on the number of tags that are present in a typical surveillance environment. Based on the matched template, we generate the natural language description of the movement as a text. For example, a typical non-pedestrian

[1] https://github.com/tiny-dnn.

Table 1 Dataset used for experimentation

Dataset	Resolution	Duration	Targets
VIRAT [14]	1080×760	30 min	30
In-House	720×480	5 h	121

target is represented by the template *"A {Color} {heavy, medium, light} {Car, Bus, Cycle, Bike, vehicle} {moving, moving high-speed, moving low-speed} from {Region Name} toward {Region Name}"* or a pedestrian can be represented as *"A Person {walking, running, loitering} from {Region Name} toward {Region Name}"*. We have used 40 different templates to describe a typical outdoor event.

4 Results

In this section, we present results of the proposed method. We have used two datasets in our experiments, namely VIRAT [14] and In-House. VIRAT is a challenging, high-quality public surveillance video of a roadside parking lot, where many pedestrians, cycles, and cars have moved around. The in-house dataset is our own recorded dataset. It is a campus video where several pedestrian and car activities around a residential zone have been recorded. Details of these datasets are given in Table 1. It may be difficult whether a description is complete or not for a given video event. Thus, we have used the guidelines proposed in [8], which involves volunteers to evaluate each description in an integer a scale between 1 to 4. Each event was rated by three volunteers and converted as a binary metric to chose the best one. All volunteers knew the regions present in the environment. Table 2 shows the results.

Quantitative evaluation of the proposed method has been performed using METEOR [3] metric which is used to evaluate machine translation results. The METEOR score is computed based on the generated text and ground truth description. Other methods such as BLEU [15] or CIDEr [18] can be used to perform similar task. Vedantam et al. [18] have shown METEOR.[2] Table 3 shows METEOR score for both datasets.

In the next subsection, we discuss qualitative and quantitative results using both datasets. First two examples show results when the proposed method is chosen as most suitable and the third one is a failed case.

[2]https://github.com/cmu-mtlab/meteor.

Table 2 Number of times a human expert chose best representation text for different method and dataset

Dataset	Number of target	Number of times choicen		
		[19]	[12]	Proposed
VIRAT [14]	30	9	6	15
In-House	121	17	26	79

Table 3 METEOR score in %, higher is better

Dataset	Number of target	METEOR		
		[19]	[12]	Proposed
VIRAT [12]	30	11	31	46
In-House	121	19	37	68

4.1 VIRAT Dataset

Ground Truth: A white car in normal speed from Roadside Zone to Corner Zone.
Video Caption [12]: A car is moving on road.
Short-term video [19]: A car is going.
Proposed: A white car with normal speed from Roadside Zone to Corner Zone.

Ground Truth: A black car with normal speed from Treeside Zone to Camera Zone.
Video Caption [12]: A car moving on road.
Short-term video [19]: A car is moving.
Proposed: A black car with normal speed from Treeside Zone to Camera Zone.

Ground Truth: A group of people loitering inside Parking Zone.
Video Caption [12]: A group of person in parking lot.
Short-term video [19]: A group of persons is moving.
Proposed: A black car with slow speed inside Parking Zone.

4.2 In-House Dataset

Ground Truth: A bike with normal speed from Residential Zone to Main Gate.
Video Caption [19]: A person riding bike.
Short video [12]: A bike is running.
Proposed: A bike is moving normal speed from Residential Zone to Main Gate.

Ground Truth: A person walking from Residential Zone to Main Gate.
Video Caption [19]: A person on the road.
Short video [12]: A person is walking.
Proposed: A person walking from Residential Zone to Main Gate.

Ground Truth: A Two persons walking from Main Building toward Main Gate.
Video Caption [19]: Walking persons.
Short video [12]: Two persons are walking.
Proposed: A bike moving slow from Main Building to Main Gate.

5 Conclusion and Future Work

This paper presents a new approach to describe surveillance events using textual descriptions. We have used hierarchical deep neural network with supervised scene knowledge. It can improve smart monitoring system, surveillance video understanding, large amount surveillance video indexing and retrieval methods. Rather than representing video sequence by a large trained actions, we mainly focus to describe a scene-specific surveillance events by tracking, classifying, and monitoring the behavior of the targets. We have observed that human experts tend to prefer the proposed textual description as the methods proposed in [12, 19] are lacking supervised scene knowledge.

There are several extensions possible. It can be applied in trajectory analysis, video summarization, abnormality detection, and surveillance report generation of long-duration surveillance video, etc.

References

1. Aradhye, H., Toderici, G., Yagnik, J.: Video2text: learning to annotate video content. In: IEEE International Conference on Data Mining Workshops, 2009. ICDMW'09, pp. 144–151. IEEE (2009)
2. Chen, X., Lawrence, Z.C.: Mind's eye: a recurrent visual representation for image caption generation. In: Proceedings of the IEEE Conference on Computer Vision and Pattern Recognition, pp. 2422–2431 (2015)
3. Denkowski, M., Lavie, A.: Meteor universal: language specific translation evaluation for any target language. In: Proceedings of the Ninth Workshop on Statistical Machine Translation, pp. 376–380 (2014)
4. Dogra, D.P., Ahmed, A., Bhaskar, H.: Smart video summarization using mealy machine-based trajectory modelling for surveillance applications. Multimed. Tools Appl. **75**(11), 6373–6401 (2016)
5. Donahue, J., Anne, H.L., Guadarrama, S., Rohrbach, M., Venugopalan, S., Saenko, K., Darrell, T.: Long-term recurrent convolutional networks for visual recognition and description. In: Proceedings of the IEEE Conference on Computer Vision and Pattern Recognition, pp. 2625–2634 (2015)
6. Guadarrama, S., Krishnamoorthy, N., Malkarnenkar, G., Venugopalan, S., Mooney, R., Darrell, T., Saenko, K.: Youtube2text: recognizing and describing arbitrary activities using semantic

hierarchies and zero-shot recognition. In: Proceedings of the IEEE International Conference on Computer Vision, pp. 2712–2719 (2013)
7. Henriques, J.F., Caseiro, R., Martins, P., Batista, J.: High-speed tracking with kernelized correlation filters. IEEE Trans. Pattern Anal. Mach. Intell. **37**(3), 583–596 (2015)
8. Hodosh, M., Young, P., Hockenmaier, J.: Framing image description as a ranking task: data, models and evaluation metrics. J. Artif. Intell. Res. **47**, 853–899 (2013)
9. Huang, H., Lu, Y., Zhang, F., Sun, S.: A multi-modal clustering method for web videos. In: International Conference on Trustworthy Computing and Services, pp. 163–169. Springer (2012)
10. Karpathy, A., Fei-Fei, L.: Deep visual-semantic alignments for generating image descriptions. In: Proceedings of the IEEE Conference on Computer Vision and Pattern Recognition, pp. 3128–3137 (2015)
11. Kiros, R., Salakhutdinov, R., Zemel, R.S.: Unifying Visual-semantic Embeddings with Multimodal Neural Language Models (2014). arXiv:1411.2539
12. Krishnamoorthy, N., Malkarnenkar, G., Mooney, R.J., Saenko, K., Guadarrama, S.: Generating natural-language video descriptions using text-mined knowledge. In: AAAI, vol. 1, p. 2 (2013)
13. Kuznetsova, P., Ordonez, V., Berg, T.L., Choi, Y.: Treetalk: composition and compression of trees for image descriptions. TACL **2**(10), 351–362 (2014)
14. Oh, S., Hoogs, A., Perera, A., Cuntoor, N., Chen, C.C., Lee, J.T., Mukherjee, S., Aggarwal, J., Lee, H., Davis, L., et al.: A large-scale benchmark dataset for event recognition in surveillance video. In: IEEE conference on Computer Vision and Pattern Recognition (CVPR), pp. 3153–3160. IEEE (2011)
15. Papineni, K., Roukos, S., Ward, T., Zhu, W.J.: Bleu: a method for automatic evaluation of machine translation. In: Proceedings of the 40th Annual Meeting on Association for Computational Linguistics, pp. 311–318. Association for Computational Linguistics (2002)
16. Rohrbach, A., Rohrbach, M., Tandon, N., Schiele, B.: A dataset for movie description. In: Proceedings of the IEEE Conference on Computer Vision and Pattern Recognition, pp. 3202–3212 (2015)
17. Rohrbach, M., Qiu, W., Titov, I., Thater, S., Pinkal, M., Schiele, B.: Translating video content to natural language descriptions. In: Proceedings of the IEEE International Conference on Computer Vision, pp. 433–440 (2013)
18. Vedantam, R., Lawrence, Zitnick, C., Parikh, D.: Cider: consensus-based image description evaluation. In: Proceedings of the IEEE Conference on Computer Vision and Pattern Recognition, pp. 4566–4575 (2015)
19. Venugopalan, S., Rohrbach, M., Donahue, J., Mooney, R., Darrell, T., Saenko, K.: Sequence to sequence-video to text. In: Proceedings of the IEEE International Conference on Computer Vision, pp. 4534–4542 (2015)
20. Venugopalan, S., Xu, H., Donahue, J., Rohrbach, M., Mooney, R., Saenko, K.: Translating Videos to Natural Language Using Deep Recurrent Neural Networks (2014). arXiv:1412.4729
21. Vinyals, O., Toshev, A., Bengio, S., Erhan, D.: Show and tell: a neural image caption generator. In: Proceedings of the IEEE Conference on Computer Vision and Pattern Recognition, pp. 3156–3164 (2015)
22. Wei, S., Zhao, Y., Zhu, Z., Liu, N.: Multimodal fusion for video searchreranking. IEEE Trans. Knowl. Data Eng. **22**(8), 1191–1199 (2010)
23. Welch, G., Bishop, G.: An introduction to the Kalman filter. In: Annual Conference Computer Graphics Interactions Technology, pp. 12–17. ACM (2001)
24. Yao, L., Torabi, A., Cho, K., Ballas, N., Pal, C., Larochelle, H., Courville, A.: Describing videos by exploiting temporal structure. In: Proceedings of the IEEE International Conference on Computer Vision, pp. 4507–4515 (2015)
25. Zivkovic, Z.: Improved adaptive Gaussian mixture model for background subtraction. In: Proceedings of the 17th International Conference on Pattern Recognition, 2004. ICPR 2004, vol. 2, pp. 28–31. IEEE (2004)

Real-Time Intrusion Detection System Using Computational Intelligence and Neural Network: Review, Analysis and Anticipated Solution of Machine Learning

Abhishek Tiwari

Abstract Today, intrusion detection system using the neural network is an interested and considerable area for the research community. The computational intelligence systems are defined on the basis of the following parameters: fault tolerance and adaptation; adaptable the requirements of make a better intrusion detection model. In this paper, provide an overview of the research progress using computational intelligence to the problem of intrusion detection. The goal of this paper summarized and compared research contributions of Intrusion detection system using computational intelligence and neural network, define existing research challenges and anticipated solution of machine learning.

Keywords Intrusion detection · Neural network · Computational intelligence

1 Introduction

Nowadays, protection of network and system via malware and attacks is a very challenging task. The requirement of intrusion detection system depends upon the intrusion recognition, data collection and preprocessing. Intrusion recognition is most vital, among these. Observe data and detection model are compared and identify the patterns of intrusion behavior either its successful or unsuccessful.

At the earlier age of intrusion detection system e focused only ho to implement the intrusion detection system, according to Denning [1] observed this major research identification in 1987. During 1980s to1990s, a hybrid expert system and statistical approaches were very much famous. Detection models were generated from the domain knowledge of security experts. Set of training data discovery using artificial intelligence and machine learning techniques. Generally, we adapt the methods for a set of training data is data classification, rule-based induction, and data-clustering. In intrusion detection problems, data are not trivial when the process of automatically

A. Tiwari (✉)
Galgotias Educational Institutions Greater Noida, Noida, UP 201306, India
e-mail: abhishektiwari.122101.cse@gmail.com

P. Chandra et al. (eds.), *Information Technology and Applied Mathematics*,
Advances in Intelligent Systems and Computing 699,
https://doi.org/10.1007/978-981-10-7590-2_11

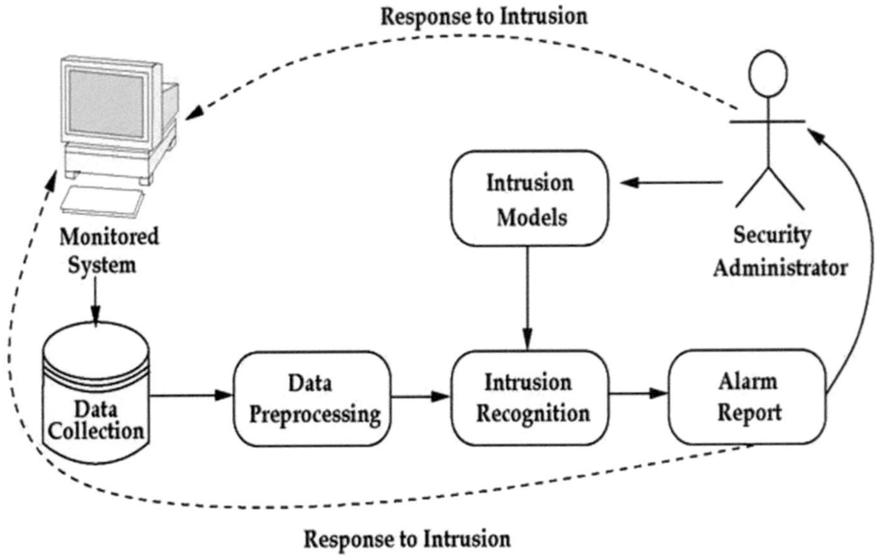

Fig. 1 Fundamental block diagram of intrusion detection system

constructing models. There are several challenges during intrusion detection system such as distinguishes criteria for normal and abnormal behavior, dynamically update the process of learning, huge network traffic. Thus, at that criteria machine learning and artificial intelligence unable to handle the solution so computational intelligence play measure role into this significance.

The goal of this review is categories is that describe the survey regarding contribution of intrusion detection model on the basis of survey, identify the existing problem and proposed solution and suggestion regarding same.

2 Related Work

2.1 Intrusion Detection

In intrusion detection system, dynamically observe the system and network regarding a legitimate user of the system. In Fig. 1 denotes solid lines for data or control flow, dashed lines for intrusive activities.

Intrusion detection system is categorized into two parts; they are misuse detection and anomaly detection.

The working strategy of misuse detection intrusions obtains data compare with predefined intrusive behavior and on the basis of these matching phenomena detect

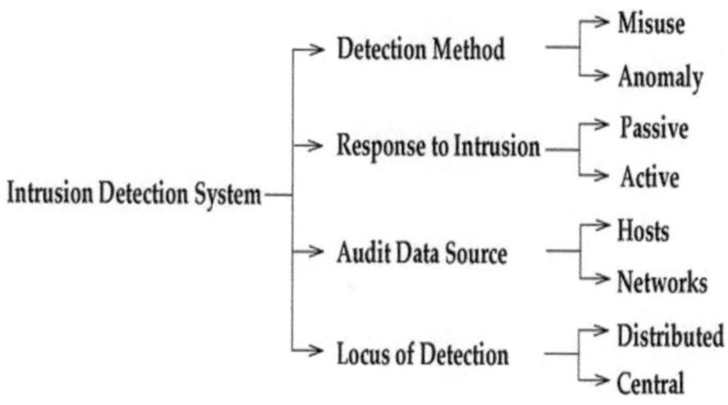

Fig. 2 Features of intrusion detection systems

the known intrusion. Due to this strength, it is highly applicable for commercial projects. Misuse detection has limitation to identify only known intrusion and is unable to identify the unknown intrusion, thus for a dynamical update of learning process used supervised learning algorithms [5].

An anomaly detection is able to identify unknown intrusion over the network and traffic, but in this working operation, dynamically updating of the learning process is a challenging task.

2.2 Computational Intelligence

The computational intelligence design of intelligent agents. It checks the illegal activities over the system and network (Fig. 2).

There are two categories of intrusion detection system, (i) misuse detection and (ii) anomaly detection. The working strategy of misuse detection is that identifies intrusions via matching obtain data with expected outcome of intrusive behavior. Thus, well-known intrusions are efficient and very low false alarm rate. Misuse detection has a limitation; it is unable to detect unknown intrusions. For the solution, its necessary update the knowledge base, via manually or automatically with the help of supervised learning algorithms [1].

3 Proposed Algorithm

There are several proposed solutions, in this section review, analysis and address the solution regarding intrusion detection problems.

Fig. 3 Diffenent categories of ANNs

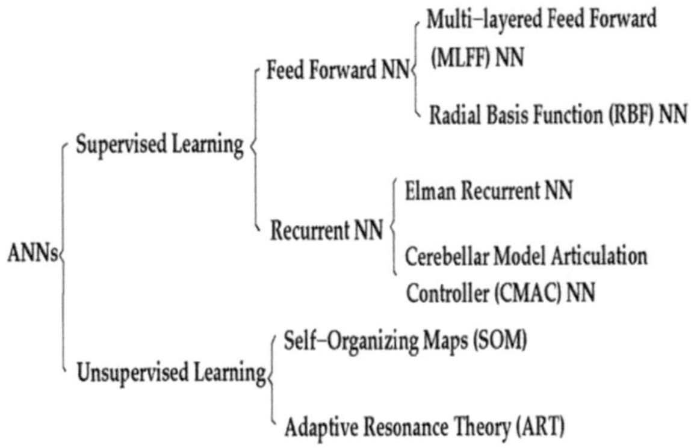

3.1 Artificial Neural Networks

In an artificial neural network, basic fundamental units are neurons. It is complex interconnected between other neurons.

The working significance is that connected neurons and corresponding weight factor sum and produces outcomes regarding fit the curve and optimize the solution.

3.1.1 Supervised Learning

Feed-forward neural networks are the simplest approach for neural network scheme.

Multilayered feed-forward (MLFF) neural networks: back-propagation multilayered feed-forward is most applicable for learning approach. Initially, earlier stages IDSs, MLFF-BP networks were applied for anomaly detection on user behavior level [9, 10].

As per research progress, user behavior parameter changes to software behavior such as system calls [11] (Fig. 3).

3.1.2 Unsupervised Learning

There are two typical There are two categories of unsupervised neural networks: Self-organizing maps and adaptive resonance theory. Using clustering algorithm approach, centralized the similar kind of data that is the cluster. Thus when any intrusion came into the system then easily recognized due to its behavior not fit into categories clusters.

Table 1 Datasets summary regarding famous IDS

Network traffic	DARPA 1999 TCPDump Files [11]	DARPA98
		DARPA99
	KDD99 Dataset [9]	KDD99
	10% KDD99 Dataset [10]	KDD99-10
Internet exploration shootout	Dataset [8]	IES
	UNIX User Dataset [9]	UNIXDS
Sequences generated by user behavior	DARPA 1998 BSM Files [5]	BSM98
	DARPA 1999 BSM Files [5]	BSM99
	University of New Mexico Dataset [6]	UNM

Self-organizing Maps

In self-organizing maps, single-layer feed-forward networks take the input and connected mechanism that is topology based on behavior similarity.

In anomaly detection schema for training purpose, self-organizing map plays a very important role [10] (Table 1).

3.2 Performance Evaluation

The intrusion detection system effective performance depends upon the protection of intrusion attacks such as confusion matrix, whose compare our actual outcomes and predict outcomes. It is natural based on the event, true negatives are true positives outcomes get depends upon normal behavior and attacks. False negatives occur when events are incorrectly predicted as normal events. Confusion matrix has four parameters regarding performance such as true-negative rate, false-positive rate, false-negative rate, and accuracy.

4 Experimental Outcomes

In this section, we simulate the proposed algorithm and find out experimental outcome. During experimental analysis of proposed algorithm, we conclude how to improve detection accuracy via ANNs (Figs. 4, 5, 6, and 7).

For the purpose of improving the detection accuracy of artificial neural network, there are following practical suggestion:

Temporal locality property: The locality of reference exists into intrusion detection behavior, so artificial neural network used this parameter and reduced the error

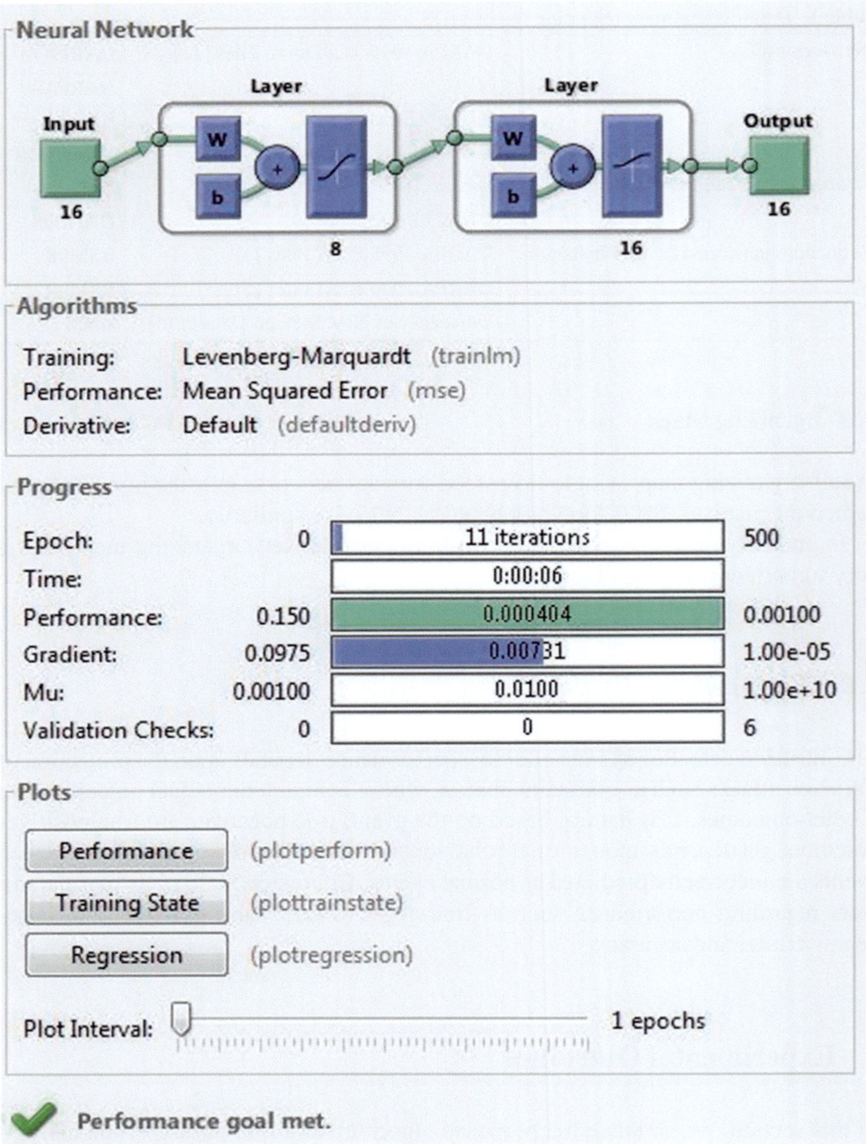

Fig. 4 Simulate two-layer neural network with respect to 16 epochs

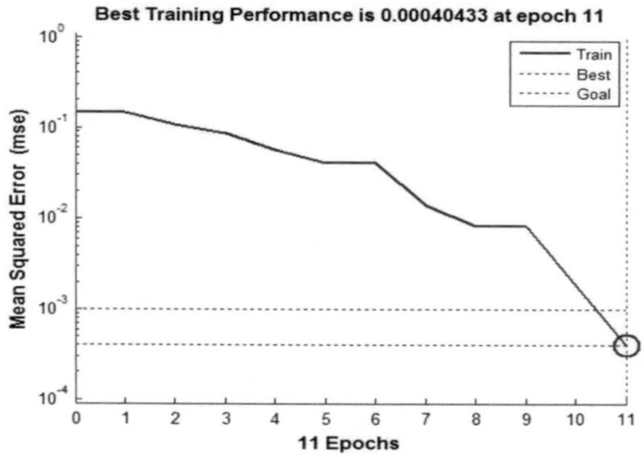

Fig. 5 Performance graph between MSE and epochs

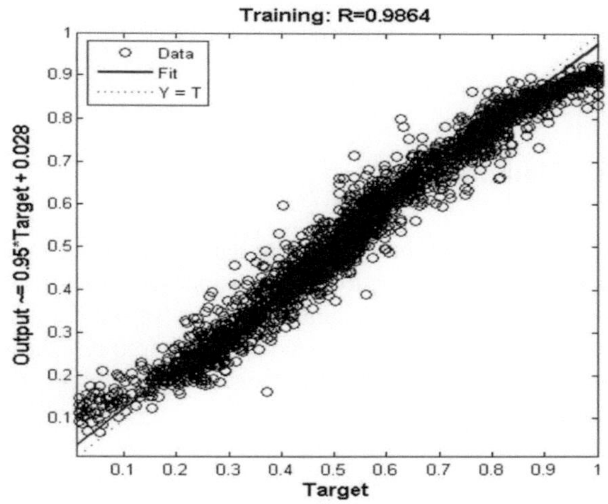

Fig. 6 Regression between target and output

regarding best-fit prediction of the curve [3]. The locality of a reference model that is based on intrusion detection behavior gives the better results.

Network structure: Intrusions are arriving at any time. Attacker attack anywhere either system-specific drive, operating system, or any application, so single-layered neural network may be unable to tackle the situation. Thus, the hierarchial structure with multilayer neural network is better for this circumstance.

Datasets and features: Neural networks only recognize whatever is fed with respect to inputs. The challenges are that when unseen intrusion came then how a system will identify. For the address, this solution, generate the random input and learn and identify the systematic pattern of distinguishing form.

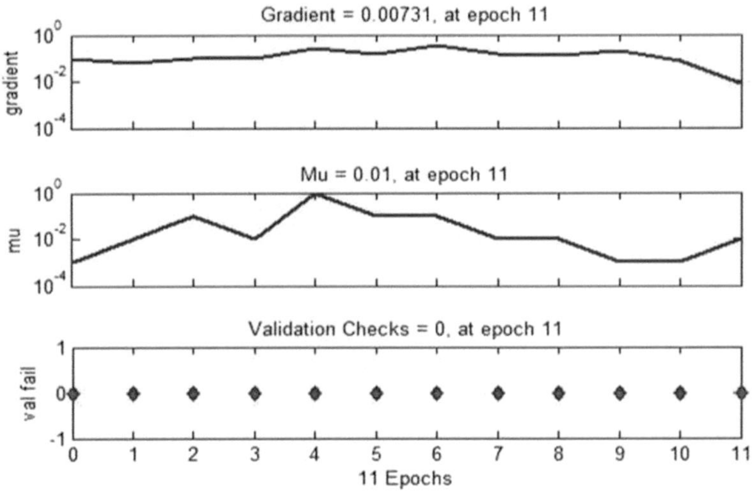

Fig. 7 Gradient with respect to epoch

5 Conclusion

In this paper presents the research progress of computational intelligence and neural network methods regarding intrusion detection systems and proposed a solution, when unseen intrusion came then how we are able to predict these intrusions. This paper also able to describe existing research direction/challenges and how learning process of intrusion detection system make better strength regarding prediction of intrusion.

References

1. Harshal, A.S., Tareek, M.P.: A comparative performance evaluation of intrusion detection based on neural network and PCA. IEEE Xplore (2015). https://doi.org/10.1109/ICCSP.2015.7322612
2. Qiuwen, C. et al.: AnRAD: a neuromorphic anomaly detection framework for massive concurrent data streams. IEEE Trans. Neural Netw. Learning Systems, 17 March 2017, https://doi.org/10.1109/tnnls.2017.2676110
3. The KDD99 Dataset. Retrieved from January 26, 2008, http://kdd.ics.uci.edu/databases/kddcup99/task.html
4. The Internet Exploration Shootout Dataset. Retrieved 2008 from http://ivpr.cs.uml.edu/shootout/network.html
5. The Unix User Dataset. Retrieved January 26, 2008, from http://kdd.ics.uci.edu/databases/UNIX_user_data/UNIX_user_data.html
6. Atencia, M., et al.: Advances in computational intelligence: selected and improved papers of the 12th international work-conference on artificial neural networks (IWANN). Neurocomputing **164**, 1–4 (2013)

7. Jabez, J., Muthukumar, B.: Intrusion Detection System (IDS): anomaly detection using outlier detection approach, international conference on intelligent computing, communication & convergence. Proc. Comput. Sci. **48**, 338–346 (2015)
8. Chen, Y.: Intelligent intrusion detection based on soft computing. IEEE Xplore (2015). https://doi.org/10.1109/icmtma.2015.145
9. Rui, Q.: Intrusion detection method research based on optimized self-buildup clustering neural network, IEEE Xplore. https://doi.org/10.1109/icia.2004.1373338(2005)
10. Muamer, N., Mohammed et al.: Intrusion Detection System Based on SVM for WLAN, Elsevier Procedia Technology (2012). https://doi.org/10.1016/j.protcy.2012.02.066
11. Aziz, A.: A similarity solution for laminar thermal boundary layer over a flat plate with a convective surface boundary condition. Commun. Nonlin. Sci. **4**(4), 1064–1068 (2009)
12. Batchelor, G.K.: The effect of Brownian motion on the bulk stress in a suspension of spherical particles. J. Fluid Mech. **83**, 97–117 (1977)
13. Chiu, H.H.: Boundary Layer Flow with Suspended Particles, p. 620. Princeton University, Report- (1962)
14. Datta, N., Mishra, S.K.: Boundary layer flow of a dusty fluid over a semi-infinite flat plate. Acta Mechanica, **42**(1–2), 71–83 (1982)
15. Makinde, O.D., Olanrewaju, P.O.: Buoyancy effects on thermal boundary layer over a vertical plate with a convective surface boundary condition. Trans. AMSE J. Fluid Eng. **132**, 1–4 (2010)
16. Mishra, S.K., Tripathy, P.K.: Mathematical and numerical modeling of two phase flow and heat transfer using non-uniform grid. Far East J. Appl. Math. **54**(2), 107–126 (2011)
17. Mishra, S.K., Tripathy, P.K.: Numerical investigation of free and forced convection two-phase flow over a wedge. Mech. Thermics **82**(1), 92–103 (2013)
18. Otterman, B.: Laminar Boundary Layer Flow of a Two Phase Suspension. Ph. D. Thesis, State University of New York at Stony Brook (1968)
19. Otterman, B., Soo, S.L.: Particulate velocity and concentration profiles for laminar flow of a suspension over a flat plate. Heat Transfer and Fluid Mechanics Institute, Monterey, California, June 10–12, Proceedings. Stanford University Press, pp: 311–322 (1970)
20. Panda, S., Misra, A., Mishra, S.K.: Flow and heat transfer in a radiative boundary layer of dusty fluid with internal heat generation/absorption over a flat plate. Appl. Math. Elixir Int. J. **84**, 33430–33439 (2015)
21. Samantara, T.N., Mishra, S.K., Panda, T.C.: Flow and heat transfer in a laminar two-phase plane wall jet. Int. J. Eng. Trends Technol. **4**(8), 3434–3447 (2013)
22. Singleton, R.E.: The Compressible Gas—Solid Particles Flow over a Semi-infinite Flat Plate. ZAMP, vol. 16, pp. 421–449 (1965)
23. Soo, S.L.: Particulate and Continuum: Multiphase Fluid Dynamics. Hemisphere Publishing Corporation, New York (1989)
24. Tripathy, P.K., Bishoyi, S.S., Mishra, S.K.: Numerical investigation of two-phase flow over a wedge. Int. J. Numeric. Methods Appl. **8**(1), 45–62 (2012)
25. Tripathy, P.K., Sahu, A.R., Mishra, S.K.: Numerical simulation of forced convection two-phase flow over an adiabatic plate. Int. J. Math. Sci. **33**(1), 1154–1159 (2012)
26. Tripathy, P.K., Samantara, T.N., Mishra, S.K.: Two-phase laminar wall jet flow with electrification of particles. Int. J. Eng. Trends Technol. **42**(9), 478–482 (2016)

A New Dual Image-Based Steganographic Scheme for Authentication and Tampered Detection Using (7, 4) Hamming Code

Partha Chowdhuri, Pabitra Pal and Biswapati Jana

Abstract Here we proposed an efficient dual image-based reversible steganographic technique using (7,4) Hamming code for image authentication and tampered detection. In this approach, the cover image is partitioned into (1×7) pixel blocks. The blocks are copied into two separate images, and three Least Significant Bits (LSB) are collected from each block separately. Then perform odd parity adjustment using Hamming code. Two symmetry keys κ and ξ are used for data embedding and stego block distribution among dual images, respectively. To embed secret bit complement data bit at κ position, then secret data bit is embedded through error creation at random positions within the block excluding the position of κ such that errors are corrected and cover image is recovered through (7,4) Hamming code. The proposed scheme has been verified through experimental results with existing scheme. The performance of this scheme is comparatively better in terms of visual quality. The experimental results showed that this new steganographic technique can be also applied for authentication and tampered detection.

Keywords Data hiding · Matrix coding · Hamming code · Correlation coefficient · Dual image · Steganography · Payload · Standard deviation

---

P. Chowdhuri (✉) · P. Pal · B. Jana
Department of Computer Science, Vidyasagar University, Midnapore 721102,
West Bengal, India
e-mail: prc.email@gmail.com

P. Pal
e-mail: pabipaltra@gmail.com

B. Jana
e-mail: biswapatijana@gmail.com

P. Chandra et al. (eds.), *Information Technology and Applied Mathematics*,
Advances in Intelligent Systems and Computing 699,
https://doi.org/10.1007/978-981-10-7590-2_12

1 Introduction

The steganography is the art of hidden communication between sender and receiver through communication media using multimedia documents [1]. The steganographic schemes are useful in many application areas to solve the problem of ownership identification, copyright protection, authentication, verification, and more. The main aims of steganographic schemes are to ensure successful data embedding and data extraction from stego media. On the other hand, data should stay hidden in stego image even if the eavesdropper tampered the stego or degrading through natural phenomenon like transmission resampling, compression or filtering, etc. Kim et al. [2] proposed data hiding through hamming code (DHHC) to hide data into halftone image. To get a syndrome value they used codeword. Using XOR operation, 4-bit secret data can be hidden within 4-bit codeword. But when individual pixel in halftone image is changed, the visual distortion is introduced. Ma et al. [3] enriched Kim's method [2] by altering the pixel pair instead of individual pixel which reduces embedding capacity by half. Using space-filling curve decomposition, Lien et al. [4] introduced a Dispersed Data Hiding technique using Hamming Code (DDHHC). Chang et al. [5] introduced a reversible data hiding technique using dual stego images. A reversible data hiding scheme using dual stego images was proposed by Lee et al. [6]. In this technique, 2 bits of secret data can be embedded by modifying only a single pixel by 1 (+ or −). After that Lee and Huang [7] proposed a dual image-based reversible data hiding scheme that eliminates the drawbacks of previous schemes. Using Hamming code, Kim and Yanh [8] would gain better qualities stego images. They achieved 48.20 dB MPSNR with embedding capacity 0.86 bpp where as Lien et al.'s scheme has MPSNR of 29.66 dB for hiding 65, 536 bits data. Jana et al. [9] developed a hamming code-based data hiding technique which one is partially reversible. The average data hiding capacity measured by bits per pixel (bpp) is 0.14 maintaining high visual quality which is higher than 59.53 dB. So far, hamming code-based steganographic scheme has been developed for grayscale images, and few literatures are available to use tampered detection and authentication, which is an emerging issue. Also, the embedding capacity and visual quality are limited, when using hamming code-based steganographic schemes.

Thus, there is a lot of scopes to investigate to improve embedding capacity and enhance visual quality through hamming code-based steganographic techniques which are still an important area of research. So, our objective is to design a secure steganographic scheme which will successfully applied for tampered detection and authentications using steganographic scheme for hidden communication with good imperceptibility and capacity.

The organization of the paper is mentioned below. Section 2 contains our proposed technique. Section 3 represents the experimental results and comparisons with existing state-of-the-art methods. The robustness and attacks of the proposed scheme are represented in Sect. 4. Finally, some conclusions are drawn in Sect. 5.

2 Proposed Steganographic Scheme

We have introduced a new steganographic technique using (7,4) Hamming code for authentication and tampered detection. The schematic diagram of data hiding and data retrieval are depicted in Fig. 1 and Fig. 2, respectively.

2.1 Data Embedding

First, we partitioned the cover image into (1×7) consecutive pixel blocks then converted into binary form and copied LSB bits into two arrays M and A. Three layers of LSB bits, i.e., LSB 1, LSB 2, and LSB 3 are collected separately and used for data embedding shown in Fig. 1. We copied LSB 1 into M and A arrays and odd parity adjustment is applied in redundant bits of both the arrays. $r1$, $r2$, and $r3$ (redundant bits) of the array M are modified depending on the number of 1 exists in the M array which are depicted in TAB 1 (Fig. 1). For an example, the $r2$ is updated by 1 if number of 1's are even in the bit positions 3, 6, and 7 of array M. Accordingly, redundant bits of A array are updated using TAB 2 of Fig. 1. Now, data embedding is started using symmetry key κ to increase security. Then data hiding started by complementing bit at κ position of array M. After that secret bit is embedded by creating an error excluding κ position. For an example, 3, 5, 6, and 7 locations are appropriate for error creation (shown in Fig. 1). Now, secret key (ω) is assigned the position of data embedded in M. The (ω) is treated as the secret key of array A. Same data embedding process is performed for the array A. After completion data embedding in the LSB 1, we apply same process for two copies of LSB 2 and LSB 3 into two arrays M and A separately. Store the pixel blocks, which are modified in stego images ($S1$ and $S2$). After that update κ by ω for the next (1×7) pixel block using $\kappa_{i+1} = ((\kappa_i \times \omega)\ mod\ 7) + 1$, where i is the number of blocks. Then continue the process and update pixel values to generate two stego images ($S1$ and $S2$). Finally, distribute pixel blocks depending upon the another key ξ (128 bits), among stego images ($S1^{'}$, $S2^{'}$). $S1^{'}$ stores the pixel from S1 and $S2^{'}$ stores the pixel from S2 when $(\xi = 1)$. Otherwise $S2^{'}$ stored the pixel from S1 and $S1^{'}$ stores the pixel from S2.

2.2 Data Extraction and Image Reconstruction

At first the receiver rearrange the block of pixels by applying the key ξ on stego images $S1^{'}$ and $S2^{'}$ shown in Fig. 2. $S1^{'}$ stores the pixel on $S1_i$ and $S2^{'}$ stores the pixel on $S2_i$ when $(\xi = 1)$. Otherwise $S2^{'}$ stored the pixel on $S1_i$ and $S1^{'}$ stores the pixel on $S2_i$. Collect LSBs of pixel blocks to perform data extraction process. LSB 1 of the first block from both stego images $S1_i$ and $S2_i$ are copied into MLSB1

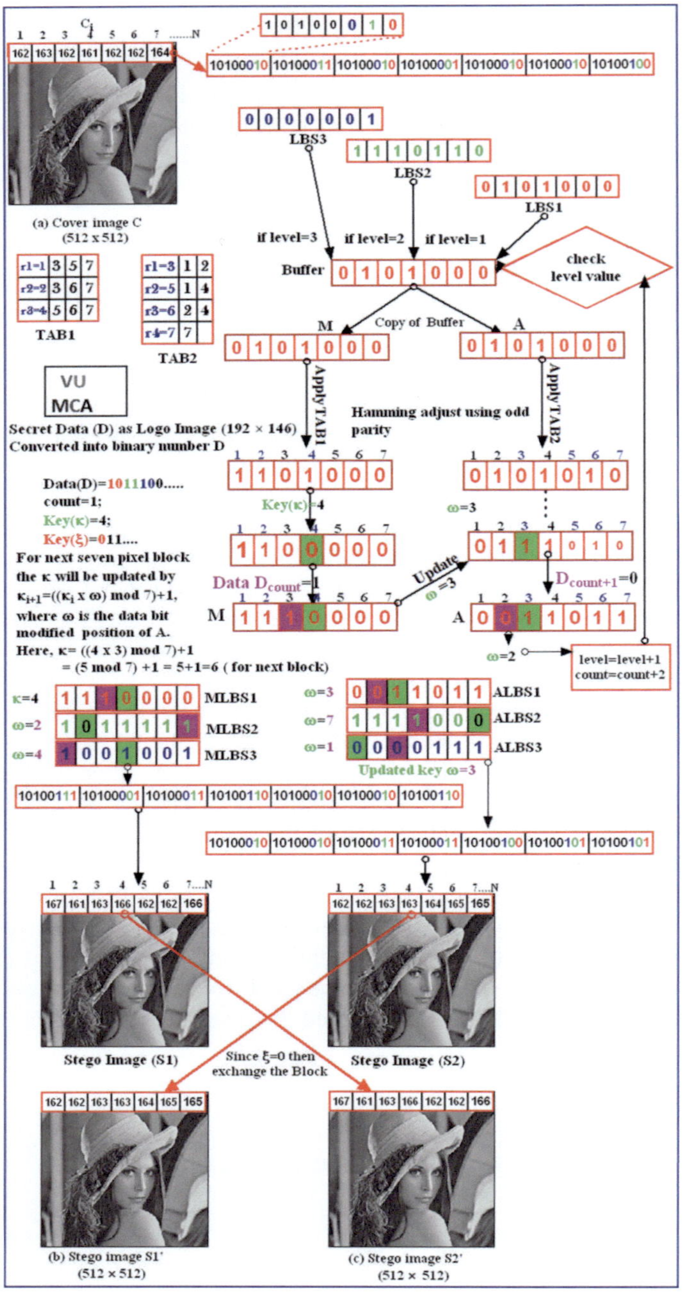

Fig. 1 Schematic diagram of hidden data embedding within dual image

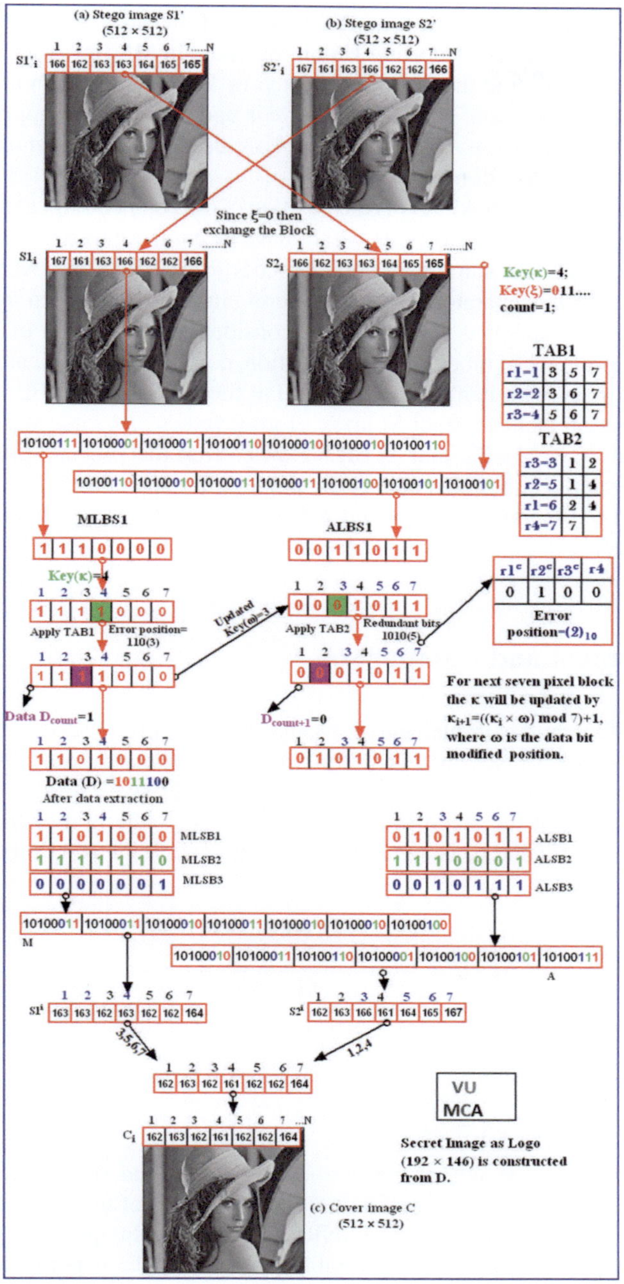

Fig. 2 Illustration of hidden data retrieval and cover image recovery

and ALSB1, respectively. Now, the κ bit position of the MLSB1 is complemented. Then error position is detected using Hamming code. In MLSB1, we use redundant bits $r1$, $r2$, and $r3$ to find the error as depicted in TAB 1 (Fig. 2). In this case error encountered at the position 3 of MLSB1, so ω is updated by 3 and apply as the ω as key for ALSB1 array. Here $r1$, $r2$, $r3$, and $r4$ are used for ALSB1 shown in TAB 2 (Fig. 2) to find error location.

The position of error in ALSB1 is the position of data embedding, and the position is again set by key ω for LSB2 bits which have to copied into two arrays MLSB2 and ALSB2. Apply the same extraction process for MLSB2 and ALSB2 and for MLSB3 and ALSB3 separately to extract secret bits. The κ has been updated using $\kappa_{i+1} = ((\kappa_i \times \omega)\ mod\ 7) + 1$, where ω is considered the position of ALSB3 and $i = 1, 2, \ldots,$ number of blocks. After extraction, compliment the bit at that position to get cover image with hamming adjusted. Put the MLSB1, MLSB2, and MLSB3 in proper position and construct M array of gray values and same way generate A array from ALSB1, ALSB2, and ALSB3. Now convert that bits into pixel value and stored on $S1^i$ and $S2^i$. Finally, collect the pixel of the positions 3, 5, 6, and 7 from $S1^i$ and pixel of the positions 1, 2, and 4 from $S2^i$ and rearrange them to construct original image.

3 Experiment and Comparison

The proposed method is implemented through Java 8. Here (512 × 512) image has been used as cover image in this study which are collected from [14] shown in Fig. 3. We have used secret message as logo image of size (192 × 146). Quality comparison between the stego image and the cover image is measured by mean square error (MSE) and peak signal-to-noise ratio ($PSNR$) using formula:

$$MSE = \frac{\sum_{i=1}^{R}\sum_{j=1}^{C}[X(i,j) - Y(i,j)]^2}{(R \times C)} \tag{1}$$

$$PSNR = 10\, log_{10} \frac{I_{max}^2}{MSE} dB \tag{2}$$

where R and C are the size of cover image ($X(i, j)$) and stego image ($Y(i, j)$), respectively, where I_{max} is the maximum intensity value of the original image and it is considered as 255. Higher the PSNR assures better image quality, and lower PSNR means poor quality of image. Table 1 represents the experimental value of average PSNR which is greater than 53 dB. The maximum embedding capacity is $6 \times (512 \times 512/7) = 2, 24, 256$ bits. The bpp in this scheme $= \frac{Totalembeddedbits}{(2 \times R \times C)} =$ 0.43. Here, we used two separate secret keys at the time of embedding phase and

Fig. 3 Images used as cover with (512 × 512) pixel

Table 1 Comparison of proposed scheme with Hamming+1, Hamming+3, Nearest Code(NC), Cao et al. Chang et al. in terms of visual quality (PSNR)

Cover image	Hamming + 1 [10]	Nearest code [11]	Hamming + 3 [8]	Cao et al. [12]	Chang et al. [13]	Proposed scheme
Lena	52.43	47.02	48.22	51.14	50.84	53.40
Baboon	53.71	–	48.18	–	50.95	53.39
Tiffany	47.46	47.03	48.20	51.15	50.11	53.14
Barbara	48.60	47.01	48.22	51.15	50.36	53.38
Aeroplane	51.61	–	48.20	–	50.77	53.40
Boat	49.37	–	48.20	–	50.39	53.41
Zelda	54.04	–	48.21	–	50.94	53.38
Pepper	47.26	47.02	48.20	51.14	50.12	53.37
Avg. PSNR	**50.56**	**47.02**	**48.20**	**51.15**	**50.56**	**53.36**

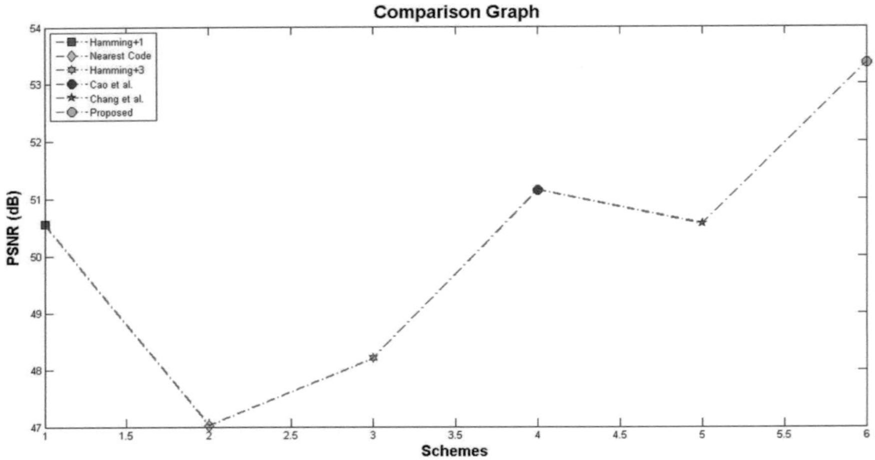

Fig. 4 Comparison graph in terms of PSNR

extraction phase, to increase the security. The comparison graph in terms of PSNR is shown in Fig. 4.

4 Robustness

4.1 Steganographic Attacks

Figure 5 depicted some experimental results of tampered detection and authentication using our proposed algorithm. Three types of experiments are presented to show the tampered detection and recovery of original image. The first one is used 20% tampered image through salt-and-pepper noise. After performing data extraction, we observed that it has been successfully detected that the stego image has been tampered and particular logo image was not present within that tampered stego image. As a result, the sender can verify the authentication of marked image. The second one is used 20% tampered through opaque. It is shown that the hidden logo image has not been recovered using data extraction and tampered cover image is recovered. The difference of SD and CC is 1.84 and 0.96, respectively. The last one, we have used tampered image through 20% cropping. Here, it also shows the recover hidden logo image and cover image, but the quality is not up to the mark. From this experiment, it is mentioned that the adversary cannot extract the logo image from the stego image through various tampering approach. But the owner can easily identify and verify the originality of the cover media. If the hidden logo is not present within any marked object, then we can easily detect that the object is tampered otherwise it

Fig. 5 Results of data extraction from tampered stego image of our proposed scheme

can be successfully verified. So this scheme may be used for image authentication, ownership identification, copyright protection, and tampered detection scheme.

4.2 Brute Force Attacks

The proposed scheme protects multimedia documents by embedding secret image (logo) using (7,4) Hamming code. The κ has been used to embed secret bits within the block. Among dual images, the distribution of stego block is done by ξ. Figure 6 illustrates how much the secret image is revealed when incorrect κ and incorrect ξ are used. Without the correct κ and correct ξ, it is impossible to reveal the secret logo. Figure 6 shows original image, logo image, attack with unknown keys κ and ξ. The result indicates an attacker can get images with noises only, when applying incorrect

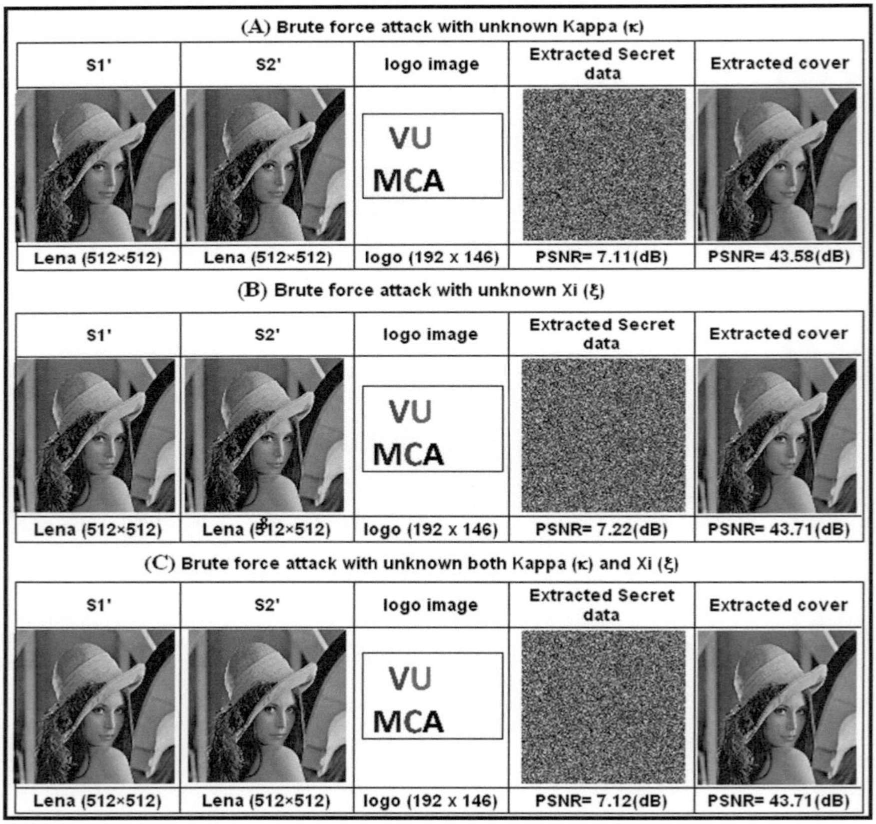

Fig. 6 Experimental results of brute force attack with unknown weighted matrix and secret key

secret key κ to reveal the logo image. But all the cases extracted cover image are as similar as the original image with good visual quality. In this context, it is mentioned that the proposed data hiding scheme is highly robust and only authorized person can extract the logo image from dual stego image. Furthermore, brute force attack can be used to reveal the logo image, which is hard for current computers to compute. The proposed technique achieves higher robustness against different attacks. Again, the logo image and the original image can be successfully recovered for dual stego images using valid secret keys κ and ξ.

5 Conclusion

In this paper, a reversible data hiding scheme based on dual images is introduced. Here, to embed confidential messages in dual image, shared secret keys are used with the help of (7,4) Hamming code. The PSNR is reasonable and higher than 53 dB. The message length is not required to retrieve the hidden message, that means the receiver can retrieve the secret message successfully, without any knowledge of the message length. Two secret keys are used in both data embedding phase and extraction phase for each blocks, and the keys for new block that guarantees security and hard to break are updated due to large key space. In this proposed scheme, data hiding based on Hamming code achieves reversibility which claims the originality of this method. Effectiveness of proposed scheme against statistical attacks is analyzed. The proposed scheme successfully detects the tampered of marked images through verification of hidden logo image. The scheme has been used for authentication, tampered detection, ownership identification, and copyright protection. Any educational institute can be used to verify their student marksheet/admit/certificate, etc., after providing their logo image within that important documents.

References

1. Johnson, N., Katzenbeisser, S.: A Survey of Steganographic Techniques, Information Hiding. Artech House, pp. 43–78 (2000)
2. Kim, C., Shin, D., Shin, D.: Data Hiding in a Halftone Image Using Hamming Code (15, 11). In: ACIIDS 2011, LNAI 6592, pp. 372–381. Springer, Berlin, Heidelberg (2011)
3. Ma, Z., Li, F., Zhang, X.: Data hiding in halftone images based on hamming code and slave pixels. J. Shanghai Univ. Nat. Sci. **19**(2), 111–115 (2013)
4. Brian, K.L., Shan-Kang, C., Wang, W.-S., King, K.P.: Dispersed data hiding using hamming code with recovery capability. In: Advances in Intelligent Systems and Computing Springer International Publishing Switzerland, vol. 329, pp. 179–187 (2015)
5. Chang, C.C., Kieu, T.D., Chou, Y.C.: Reversible data hiding scheme using two steganographic images. In: Proceedings of IEEE Region 10 International Conference (TENCON), pp. 1–4, (2007)

6. Lee, C.F., Wang, K.H., Chang, C.C., Huang, Y.L.: A reversible data hiding scheme based on dual steganographic images. In: Proceedings of the Third International Conference on Ubiquitous Information Management and Communication, pp. 228–237 (2009)
7. Lee, C.F., Huang, Y.L.: Reversible Data Hiding Scheme Based on Dual Stegano-Images Using Orientation Combinations, Telecommunication Systems, vol. 52, No. 4, pp. 2237–2247 (2013)
8. Kim, C., Yang, C.N.: Improving Data Hiding Capacity Based on Hamming Code pp. 697–706. Springer, Netherlands (2014)
9. Jana, B., Giri, D., Mondal, S.K.: Partial reversible data hiding scheme using (7, 4) hamming code. Multimed. Tools Appl. 1–16 (2016)
10. Zhang, W., Wang, S., Zhang, X.: Improving embedding efficiency of covering codes for applications in steganography. IEEE Commun. Lett. **11**(8), 680–682, (2007)
11. Chang, C.C., Chou, Y.C.: Using nearest covering codes to embed secret information in grayscale images. In: Proceedings of the 2nd International Conference on Ubiquitous Information Management and Communication, pp. 315–320. ACM (2008, Jan)
12. Cao, Z., Yin, Z., Hu, H., Gao, X., Wang, L.: High capacity data hiding scheme based on (7, 4) Hamming code. SpringerPlus **5**(1), 175 (2016)
13. Chang, C.C., Kieu, T.D., Chou, Y.C.: A high payload steganographic scheme based on (7, 4) hamming code for digital images. In: 2008 International Symposium on Electronic Commerce and Security, pp. 16–21. IEEE (2008, Aug)
14. University of Southern California, The USC-SIPI Image Database. http://sipi.usc.edu/database/database.php

DFA-Based Online Bangla Character Recognition

Shibaprasad Sen, Dwaipayan Shaoo, Mridul Mitra, Ram Sarkar and Kaushik Roy

Abstract In the present experiment, we have investigated the effectiveness of two handcrafted feature extraction techniques for the recognition of constituent strokes of online handwritten Bangla character samples. These techniques estimate local and global shape information from a stroke sample. Combined feature vector is fed to Multi-Layer Perceptron (MLP)-based classifier for stroke recognition purpose. We have achieved 91.27% recognition accuracy over test set. In the current experiment, total size of the stroke database is 32,534. Among the samples, 30% of the entire strokes are considered as test set and rest are used to train the recognition model. Afterward, we have implemented a Deterministic Finite Automata (DFA)-based character recognition system from the recognized strokes. Final outcome of the system is satisfactory considering the stroke variation orders while writing Bangla characters.

Keywords Bangla script · Stroke recognition · Online character recognition
Zone-based path traversal feature · DFA-based recognition

S. Sen (✉) · D. Shaoo · M. Mitra
Future Institute of Engineering and Management, Kolkata, India
e-mail: shibubiet@gmail.com

D. Shaoo
e-mail: dwaipayanshaoo18@gmail.com

M. Mitra
e-mail: mitra.m150495@gmail.com

R. Sarkar
Jadavpur University, Kolkata, India
e-mail: raamsarkar@gmail.com

K. Roy
West Bengal State University, Barasat, India
e-mail: kaushik.mrgr@gmail.com

P. Chandra et al. (eds.), *Information Technology and Applied Mathematics*,
Advances in Intelligent Systems and Computing 699,
https://doi.org/10.1007/978-981-10-7590-2_13

1 Introduction

Online handwriting recognition has now established a solid footprint in pattern recognition domain due to the advantages as follows: (i) Data are collected through some online devices where people write information almost the way they write with normal pen and paper, which in turn saves time and minimizes the chances of mistyping that might occur if a keyboard is used, (ii) saved information is almost noise free, and (iii) information is stored as a ordered sequence of pixels (x-coordinate, y-coordinate, and pen-up/pen-down status). These information are very much helpful to the researchers who work on online character recognition applications. For example, this stroke order information is required while trying to reconstruct the characters/texts in a stroke-based recognition approach.

In the literature, few researches for Gurumukhi [1–6], Devanagari [7–11], and English [12–14] scripts are available, but in case of Bangla script, the presence of little research publications silently describe the need of more research in this domain.

Roy in [15] and Roy et al. in [16] have considered stroke-level sequential and ordered information as feature vector for stroke and character recognitions. In stroke-based recognition technique, Roy has constructed characters from the recognized strokes by matching stored stroke sequences in the database. Parui et al. in [17] have manually grouped the strokes into 54 classes. They have used sub-stroke-level features and designed one hidden Markov model (HMM) for individual stroke class. After stroke recognition, characters are formed with the help of 50 look-up tables. In [18–20], authors have recognized the character in a holistic approach (irrespective of number of strokes in it) by extracting shape-based and topological features. Ghosh in [21] has reported the recognition of strokes by considering direction of writing, curvature, curliness, inclination, etc. Sen et al. in [22] have used distance-based feature for the same, and then a rule base is generated for the construction of character from the recognized strokes.

After analyzing the issues mentioned in the said articles, it can be concluded that researchers are trying to develop some good-quality feature sets for stroke/character/sub-stroke recognition. Except the works reported in [15, 22], most researchers have tried to recognize the character in a holistic way (as an single input). The limitation of recognizing characters in a holistic way is that after a certain level it becomes very difficult to enhance the accuracy of recognition module. This happens because of strong structural similarity in few of the characters in the alphabet set. In contrary, the key advantage of stroke-based character recognition is that the wrong stroke-level prediction can be altered by using some intelligent techniques, which in turn help to increase the accuracy of the overall recognition module. It is to be noted that Bangla script recognition is challenging because of the presence of large alphabet set. The variation in number and order of strokes to write same character makes the thing more complex [15]. Hence, as the first step, in this present work, we have estimated relatively powerful feature set for stroke recognition. After recognizing the component strokes, we have then followed a DFA-based approach to reconstruct the characters more accurately.

2 Stroke Database Generation

For the current experiment, 100 distinct individuals were requested to write 300 samples for each handwritten character. Diverse individuals were selected considering their age, gender, qualification, etc. In the data collection process, we had not set any strict instruction to them. After analyzing the extracted strokes, we have build the stroke database by following the approach as mentioned in [22] (sample is shown in Fig. 1). Strokes cannot be collected separately, hence these are extracted from the parent character. Please note, a stroke is a sequence of pixels p_t collected in between one pen-down and one pen-up status. p_t describes pixel point having x-coordinate (x_t), y-coordinate (y_t), and pen-up/pen-down information. For naming convention of the strokes, rules mentioned in [22] have been followed. Fifty different characters are present in Bangla script; hence, the current database comprises 15000 samples in total (300 samples per character).

Usually, characters are written by using single/multiple strokes. After observing the database carefully, it was noted that from one to six strokes have been used to write a character. As evidences, এ and ও are normally written by using a single stroke while on an average four strokes have been used write উ. Whereas, at most six strokes were employed to write আ. Considering all instances, it is estimated that average number of strokes used to write a character is 2.2. Therefore, approximately 110 strokes would have been required to represent Bangla basic character set. But during experiment, it has been noticed that only 52 strokes are sufficient to represent all the basic characters as few characters share some basic strokes.

Bangla Stroke	Nomenclature	Bangla Stroke	Nomenclature	Bangla Stroke	Nomenclature	Bangla Stroke	Nomenclature
৩	3_	এ	E_	জ	JA_	[illegible]	PA2
[illegible]	_AA	ও	O_	[illegible]	EN_	প	PA3
—	MTR	ক	KA_	[illegible]	DDH	ফ	PHA
।	_A_	১	1_	ট	T_	ভ	BHA
[illegible]	AAA	খ	KH_	ঠ	TTA	ম	MA_
২	2_	গ	G_	[illegible]	TT_	য	YA_
[illegible]	I_	[illegible]	G1_	ণ	MN_	০	0_
[illegible]	II_	[illegible]	G2_	[illegible]	MN1	ল	LA_
৬	6_	ঘ	GH_	থ	THA	শ	SA_
[illegible]	UU_	ঙ	UN_	দ	DA_	—	MSA
ধ	DHA	চ	C_	[illegible]	DH1	স	S_
ব	BA_	ছ	CH_	ন	NA_	[illegible]	S1_
[illegible]	RI_	[illegible]	JA1	[illegible]	PA1	ৎ	KT_

Fig. 1 Basic strokes used to write Bangla character set

3 Pre-processing and Feature Extraction

After extracting strokes from the sample characters in the pre-processing step, at first duplicate points are removed. The contribution of pixels to write a stroke varies significantly due to speed variation while writing. To cope with this, the constituting points of the stroke are re-sampled to generate a modified ordered sequence of 120 points that are at almost equal distance apart and finally, those points are scaled to a window size of 512 × 512.

3.1 Distance-Based Feature

Distance-based features are useful for both stroke and character recognitions as mentioned in [22, 23]. When this is used for stroke recognition, at first, each stroke is segmented to certain number of parts (say, M). As each part is represented by a pair of points, the same is divided into M+1 almost equidistant points p_k (k ranges from 1 to M+1). Feature values are estimated by the distances measured from each point to every other points as shown in Fig. 2. Existence of some nearly similar shape structures has been observed during statistical analysis of the stroke database considered here. In order to differentiate such similar shape patterns, some local information are estimated to produce discriminating features for the recognition purpose. As selecting the optimal number of segmentation of the stroke is always challenging, to avoid the complexity, we have used the work [22, 23] where authors have divided each stroke into 16 segments. For details of distance-based feature extraction procedure, see [22].

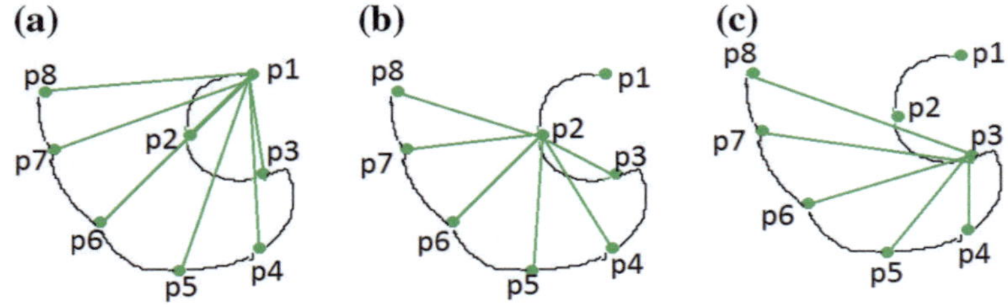

Fig. 2 Calculation of distances between segmented points

For a stroke pattern with M number of segments, this feature extraction method produces M*(M+1)/2 number of features (i.e., the distance values). Thereby a feature vector of size 136 is generated because here strokes are segmented into 16 parts.

3.2 Zone-Based Path Traversal (ZPT) Feature

In this procedure, we have divided a stroke image into 8×8 rectangular blocks. According to the writing pattern of the stroke, these blocks may have varied number of pixel points. For the extraction of ZPT feature, we have marked each block with a distinct index value as shown in Fig. 3. Then we have traversed the pixels according to the writing pattern and tried to find the longest segments within different blocks (e.g., AB, BC, CD). Please note that index of a block is set to track the positional information of the segment and by segment we mean the continuous portion/contribution of a stroke sample in a block. From the figure, it can be noticed that a particular block can have more than one such segments (due to inherent shape characteristic of a stroke). We have measured the contributions of all the segments for each block by considering the facts:

If any segment does not contain end point of a stroke and the pattern passes through say, block 1 to block 2, then, the contribution of the segment lies in block 1 is set to $((n-1) * index\_value(block1)) + index\_value(block2)$. Otherwise if this is the last segment within block 1, then the contribution is set as $(n * index\_value(block1))$. Here n denotes number of pixels within that segment. We have assumed that a pixel p_i contributes the index value of the block it resides iff the next pixel p_j lies within the same block. Otherwise, if the next pixel p_j lies in different block (say 2), then p_i contributes the index value of the block 2.

Zero-valued block implies that it does not contain any pixel. As the index value of blocks is known, hence, contributed value of any arbitrary block not only reveals the pixel density of that block along with the pixels that contribute in the neighboring blocks but also this feature extraction technique helps to identify the path (sequence of blocks) through which stroke pattern passes. The transition between blocks is depicted in Fig. 3 (pixel color depends on the index value of corresponding blocks). In the present work, block-wise computed contributions made by all the segments of a writing pattern have served as feature values. As a stroke is divided into 8×8 number of blocks, this procedure yields a 64-element feature vector. It is to be noted that this procedure tries to acquire the shape information of the stroke.

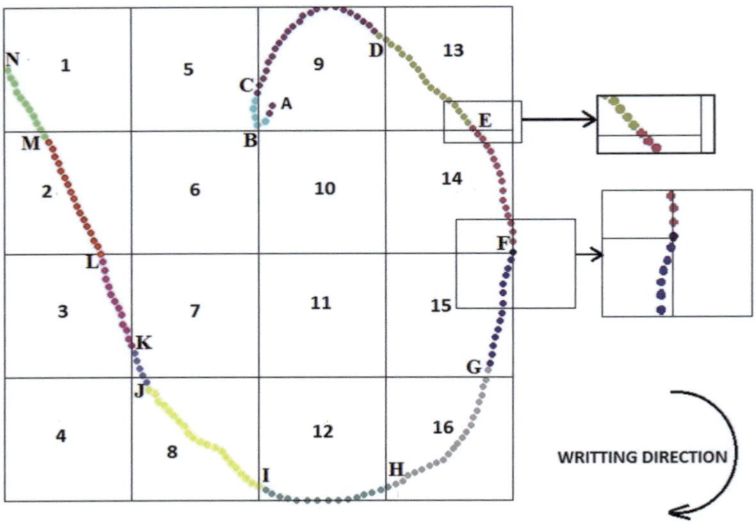

Fig. 3 Pictorial description of ZPT feature computation

4 Recognition

4.1 Stroke Recognition

The combination of distance-based and ZPT-based feature vectors produces an 200 (i.e., $136 + 64$) element feature vector which is fed to MLP for stroke recognition and achieved 91.27% classification accuracy. A 2-layered MLP (200–98–52) has been designed for classification purpose. The input, output, and hidden layer neurons are set to 200, 52, and 98, respectively. Values of acceleration factor and learning rate of the Back Propagation (BP) algorithm have been finalized after few trial runs.

4.2 Character Formation

It has been mentioned earlier that there are few strokes that are used to write various/multiple character samples (see Table 1 in [15] for detail). From this table, we have observed that few strokes (e.g., BA_, 3__, 6__) describe the major part of the character, while some strokes help to form different characters as a minor part of the character (e.g., 0__, _A_). Considering these facts, we have categorize the strokes into three groups:

Table 1 Comparison of the present work with some past works

Reference	Used features	Feature dimension	Database size	Recognition accuracy (in %)
Roy [15]	Structural + Point Based	105	21372	88.23
Sen et al. [22]	Distance Based	136	32534	89.39
Ghosh [21]	Structural + Directional	360	4900 (test samples)	87.48
Present Work	**Distance based + ZPT**	**200**	**32,534**	**91.27**

Major Stroke: It is the major portion of the character (e.g., '6__' is the major stroke of 'ড').

Minor Stroke: It contains minimal part of the character (e.g., 'I__' is the minor stroke of 'উ').

Matra: A horizontal straight line goes through the upper portion of a character (e.g., 'ড' has the Matra stroke 'MTR').

In the present work, we have analyzed all the strokes that are used to write Bangla basic characters. From this analysis, we have prepared a list of strokes that act as major parts for the construction of multiple characters (major stroke list). After tracking the major stroke of a character sample, we have rearranged the stroke sequence by placing the major stroke and Matra at the start and end positions, respectively, to minimize the stroke order variation. We have started with a node representing a major stroke and constructed a DFA to describe different paths for all possible characters that can be derived from this major stroke. Figure 4 illustrates on such DFA to recognize the characters 'ড', 'ডা' and 'ড়' from major stroke '৬'.

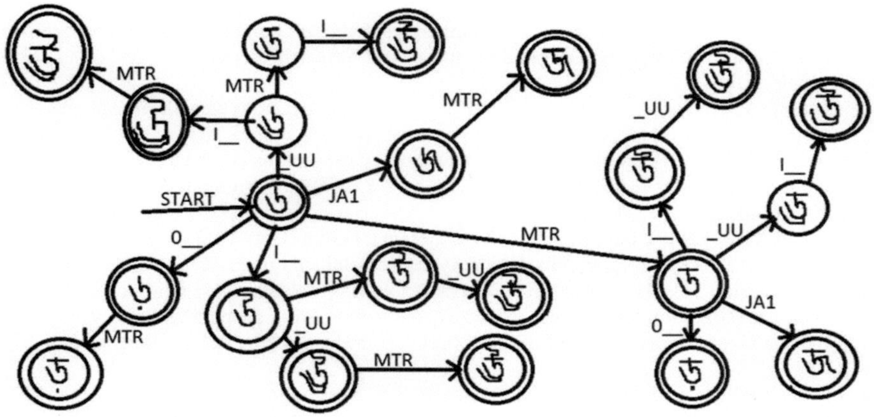

Fig. 4 Formation of different characters from major stroke '৬'

This major stroke list is important for the construction of DFAs because for each such major stroke, a separate DFA needs to be formed and most of the symbols belonging to Bangla basic alphabet set contain a major stroke. Please note major strokes are the starting node in the DFA. There are also few characters that may not contain any major stroke, but each of them contains at least one such stroke that can distinguish the character from others and treated them as starting node of DFA for that character (e.g., 'গ' can be written by using strokes $G1\_$ and $G2\_$ mentioned in Fig. 1 and both the strokes are minor strokes, but $G1\_$ can only be present character 'গ').

Table 1 records the performance comparison of the present schema with some recent works [15, 21, 22]. Experiment mentioned in [22] has been performed over the same database. The other two procedures have been mentioned in the table for the quick overview about the performances in this domain.

5 Conclusion

In this present experiment, we have focused on designing stroke recognition module by adopting two shape-based feature vectors, viz. distance based and ZPT. Though we have minimized the stroke order variation by adjusting the stroke sequence and implemented suitable DFA-based character reconstruction module, the performance of the system heavily depends on stroke recognition accuracy. Stroke misclassification leads to wrong character prediction. Thereby in future, we will plan to consider top-n choices of the stroke recognition results to find out most acceptable stroke sequence from therein that would help in better character prediction.

References

1. Sharma, A., Kumar, R., Sharma, R.K.: Online handwritten Gurmukhi character recognition using elastic matching. In: Congress on Image and Signal Processing, pp. 391–396. IEEE (2008)
2. Sharma, A., Kumar, R., Sharma, R.K.: HMM-based online handwritten Gurmukhi character recognition. Int. J. Mach. Graph. Vis. **19**, 439–440 (2010)
3. Bawa, R.K., Rani, R.: A preprocessing technique for recognition of online handwritten Gurmukhi numerals. In: IEEE International Conference on High Performance Architecture and Grid Computing, pp. 275–281 (2011)
4. Sachan, M.K., Lehal Singh, G., Jain, V.K.: A system for online Gurmukhi script recognition. In: Intrernational Conference on Information Systems for Indian Languages, pp. 299–300 (2011)
5. Swethalakshmi, H., Jayaraman, A., Chakravarthy, V.S., Sekhar, C.C.: On-line handwritten character recognition for Devanagari and Telugu scripts using support vector machines. In: Proceedings of International Workshop on Frontiers in Handwriting Recognition, pp. 367–372 (2006)
6. Swethalakshmi, H., Sekhar, C.C., Chakravarthy, V.S.: Spatio structural features for recognition of online handwritten characters in Devanagari and Tamil scripts. In: Proceedings of International Conference on Artificial Neural Networks, pp. 230–239 (2007)

7. Santosh, K.C., Iwata, J.: Stroke-based cursive character recognition. Xiaoqing Ding, INTECH (2012)
8. Santosh, K.C., Nattee, C., Lamiroy, B.: Relative positioning of stroke-based clustering: a new approach to online handwritten devanagari character recognition. Int. J. Image Graph. **12**(2), 25 (2012)
9. Connell, S.D., Sinha, R.M.K., Jain, A.K.: Recognition of unconstrained online Devanagari characters. In: 5th International Conference on Pattern Recognition, pp. 368–371 (2000)
10. Joshi, N., Sita, G., Ramakrishnan, A.G., Deepu, V.: Machine recognition of online handwritten Devanagari characters. In: Proceedings of International Conference on Document Analysis and Recognition, pp. 1156–1160 (2005)
11. Santosh, K.C., Wendlingc, L.: Character recognition based on non-linear multi-projection profiles measure. Int. J. Front. Comput. Sci. **9**(5), 678–690 (2015)
12. Tappert, C.C., Suen, C.Y., Wakahara, T.: The state of online handwriting recognition. In: Transaction on Pattern Analysis and Machine Intelligence, pp. 787–807. IEEE (1990)
13. Farha, M., Srinivasa, G., Ashwini, A.J., Hemant, K.: Online handwritten character recognition. Int. J. Comput. Sci. 30–36 (2013)
14. Alvaro, F., Zanibbi, R.: A shape-based layout descriptor for classifying spatial relationships in handwritten math. In: ACM Symposium on Document Engineering, pp. 123–126 (2013)
15. Roy, K.: Stroke-database design for online handwriting recognition in Bangla. Int. J. Modern Eng. Res. 2534–2540 (2012)
16. Roy, K., Sharma, N., Pal, U.: Online Bangla handwriting recognition system. In: International Conference on Advances in Pattern Recognition, pp. 117–122 (2007)
17. Parui, S.K., Guin, K., Bhattacharya, U., Chaudhuri, B.B.: Online handwritten Bangla character recognition using HMM. In: International Conference on Pattern Recognition, pp. 1–4 (2008)
18. Bhattacharya, U., Gupta, B.K., Parui, S.K.: Direction code based features for recognition of online handwritten characters of Bangla. In: International Conference on Document Analysis and Recognition, pp. 58–62 (2007)
19. Sen, S.P., Sarkar, R., Roy, K., Hori, N.: Recognize online handwritten Bangla characters using hausdorff distance based feature. In: 5th International Conference on Frontiers in Intelligent Computing: Theory and Application, pp. 541–549 (2017)
20. Sen, S.P., Mitra, M., Chowdhury, S., Sarkar, R., Roy, K.: Quad-tree based image segmentation and feature extraction to recognize online handwritten Bangla characters. In: 7th IAPR TC3 Workshop on Artificial Neural Networks in Pattern Recognition, Ulm, Germany, pp. 246–256 (2016)
21. Ghosh, R.: A novel feature extraction approach for online Bengali and Devanagari character recognition. In: International Conference on Signal Processing and Integrated Networks, pp. 483–488 (2015)
22. Sen, S.P., Sarkar, R., Roy, K.: An approach to stroke-based online handwritten Bangla character recognition. In: 3rd International Doctoral Symposium on Applied Computation and Security Systems (2016)
23. Sen, S.P., Sarkar, R., Roy, K.: A simple and effective technique for online handwritten Bangla character recognition. In: 4th International Conference on Frontiers in Intelligent Computing: Theory and Application, pp. 201–209 (2015)

Part II
Applied Mathematics

An Integrated Imperfect Production–Inventory Model with Optimal Vendor Investment and Backorder Price Discount

Anindita Mukherjee, Oshmita Dey and B. C. Giri

Abstract In this article, an integrated single-vendor single-buyer imperfect production–inventory model in which the vendor makes investment for process quality improvement and the buyer offers price discounts for backorders is studied. It is assumed that the buyer follows a continuous review policy with lot-size-dependent lead-time and a mixture of backorders and lost sales. Under the n-shipment policy, the expected annual total cost of the integrated system is derived. An algorithm is developed to determine numerically the optimal decisions of the model. A numerical example is taken to illustrate the developed model and to examine the sensitivity of the key parameters of the model. Some managerial insights are also provided.

Keywords Inventory · Integrated model · Defective items · Backorder price discount

1 Introduction

The problem of a single-vendor single-buyer inventory model and its various extensions have been an area of interest among the practitioners of operations research for quite some time, and a significant amount of literature is available in this regard. For instance, Goyal in [7] was among the first researchers to investigate an integrated inventory model for a single-vendor single-buyer system. Banerjee in [1] generalized the previous model and presented a joint economic lot-size model with the vendor producing items on a lot-for-lot basis. Goyal in [8] further extended this model by relaxing the assumptions of the lot-for-lot policy. Ha and Kim in [10] further generalized this model and developed an integrated lot-splitting model to

A. Mukherjee · B. C. Giri
Jadavpur University, Kolkata 700032, West Bengal, India
e-mail: bcgiri.jumath@gmail.com

O. Dey (✉)
Techno India University, Kolkata 700091, West Bengal, India
e-mail: oshmi_kgp@yahoo.co.uk

P. Chandra et al. (eds.), *Information Technology and Applied Mathematics*,
Advances in Intelligent Systems and Computing 699,
https://doi.org/10.1007/978-981-10-7590-2_14

facilitate multiple shipment in small lots. Hill in [11] proposed a more generalized batching and shipment policy. Pa and Yang in [21] extended the model developed by Goyal in [8] by considering lead-time to be a decision variable. Pa and Yang in [24] developed a model where deterministic variable lead-time and quality improvement were considered. Ben-Daya and Hariga in [3] extended the literature on single-vendor single-buyer integrated inventory model by assuming stochastic demand and thereby allowing shortages. Since then a lot of researchers, viz. Hsiao [12], Glock [5, 6], have also made significant contributions in this area. However, in most of these cases, process quality and its effect on the production shipment schedule from the vendor to the buyer were not considered. But, in real-life situations, the production process is generally not perfect. And assuming it to be so is an unrealistic restriction imposed on the model. Porteus [22] was the first to introduce the concept of process quality control and introduced the logarithmic investment function for this purpose. Rosenblatt and Lee in [23] extended this model to the EPQ model. Since then the concept of process quality improvement has been a focal area of interest among researchers. For instance, Zhang and Gerchak in [25] considered a joint lot sizing and shipment policy where the percentage of defectives units is assumed to be random, and that at the end of the screening period, the defective items are returned as a single batch. Ben-Daya and Hariga in [2] developed an economic lot scheduling problem assuming imperfect production process. A practical approach to the EPQ model with imperfect items was proposed by Goyal, Cardenas-Barron in [9]. Huang in [13] developed an integrated vendor–buyer inventory model for items with imperfect quality, where it is assumed that the number of defective items follows a given distribution. Ouyang et al. in [19] also investigated an integrated model with imperfect production process. Lin in [18] considered an integrated supply chain inventory model with imperfect quality items, controllable lead-time, and distribution-free demand. Recently, Dey and Giri in [4] extended existing literature on integrated models by considering vendor investment for process quality improvement in a stochastic integrated inventory model, thereby allowing the percentage of defective items produced to be a control parameter. However, in this model, the shortages were assumed to be completely backlogged. However, it is seen in most cases that, when shortage occurs, while some customers may be willing to wait for their order to arrive at a later stage (backorders), some may be unwilling to do so and may wish to take their orders elsewhere (lost sales). But, incurring lost sales reflect poorly on the customer's goodwill. One way of preventing the percentage of lost sales is to offer a backorder price discount to motivate customers to wait for their orders to arrive with the next batch instead of taking them elsewhere. That is, the buyer may offer the backordered items at a discounted price so as to try and increase the proportion of backorders vis-a-vis lost sales. For instance, an integrated supply chain vendor–buyer model was proposed by Lin [17] where backorder price discount and effective investment required to reduce the ordering cost were considered. Jaggi and Arneja in [14] investigated the periodic review inventory model with backorder price discounts where shortages were partially backlogged. Pal and Chandra in [20] studied a periodic review inventory model with stock-dependent demand, permissible delay in payment, and backorder price discount. Ahamed in [15] developed the lot-size decision for stochastic vendor–buyer system with quantity discount and partial backorder. Jindal and Solanki in [16] studied an integrated supply chain inventory model with quality improvement involving controllable lead-time and backorder price discounts.

Keeping these various issues in mind, an attempt is made to extend existing literature on single-vendor single-buyer integrated inventory models by developing an integrated inventory model with optimal vendor investment and backorder price discount. The model is also investigated under both constant and lot-size-dependent lead-time.

The rest of the paper is organized as follows: In Sect. 2, the proposed model is formulated, and the solution procedure is outlined in Sect. 3. Section 4 illustrates the developed model with the help of numerical examples. The paper is concluded in Sect. 5 with some remarks and future research directions.

2 Model Development

2.1 Notations and Assumptions

The following notations are used for developing the proposed model:

D: Expected demand rate for non-defective items
P: Production rate for the vendor ($P = 1/p$)
A: Ordering cost per order for the buyer
F: Transportation cost per delivery
B: Setup cost for the vendor
L: Lead-time
Q: Order quantity
h_v: Holding cost per item per unit time for the vendor
h_{b1}: Holding cost for defective items per unit time for the buyer
h_{b2}: Holding cost for non-defective items per unit time for the buyer
n: The number of shipments per production run from the vendor to the buyer
r: Reorder point
s: Screening cost per unit item for the buyer
x: Buyer's screening rate
w: Warranty cost per unit defective item for the vendor
k: Safety stock factor
y: Percentage of defective items produced
η: Fractional annual opportunity cost
δ: Percentage decrease in defective items per dollar increase in investment
β: Fraction of the shortage that will be backordered at the buyer's end ($0 \leq \beta < 1$)
β_0: Upper bound of the backorder ratio, ($0 \leq \beta \leq \beta_0 \leq 1$)
π_x: Unit backorder price discount

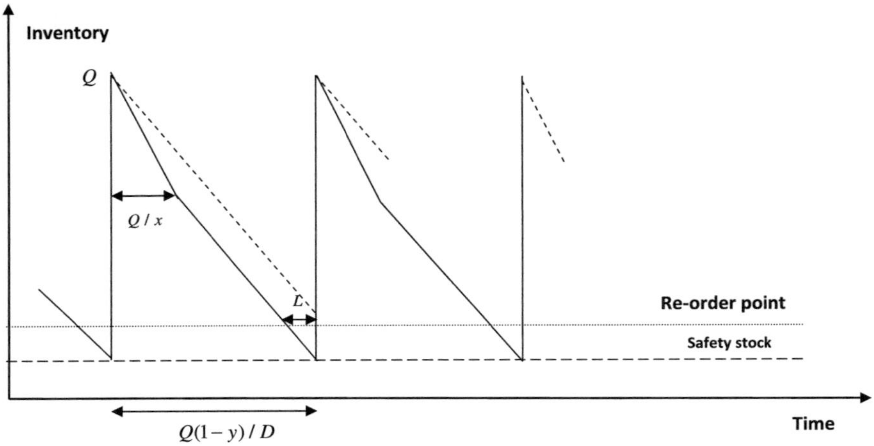

Fig. 1 Inventory of the buyer

π_0: Marginal profit per unit
X: Lead-time demand.

The model is based on the following assumptions (Figs. 1 and 2):

- A single vendor provides single type of item to a single buyer.
- Demand per unit time is normally distributed with mean D and standard deviation σ.
- The buyer orders quantity of nQ items to the vendor which then the vendor produces at one go and delivers in n equal-sized shipments of Q items each to the

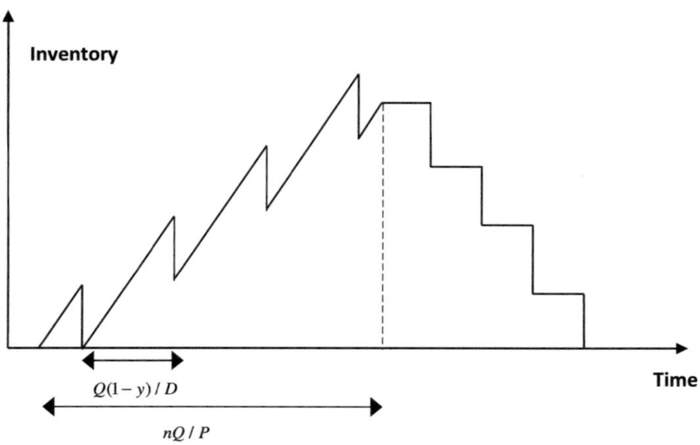

Fig. 2 Inventory of the vendor

buyer. The production rate of non-defective items is assumed to be greater than the demand rate, i.e., $P(1-y) > D$.

- The buyer follows the (Q, r) continuous review policy with constant lead-time and partial backlogging. The case of lot-size-dependent lead-time is also investigated.
- Percentage of backorders is β.
- In case of shortage, the buyer provides a price discount to the customers to encourage them to wait for their orders to arrive with the next batch.
- Lead-time L is constant, and the lead-time demand is normally distributed with mean DL and standard deviation $\sigma\sqrt{L}$.
- The reorder point r = expected demand during lead-time + safety stock (SS), i.e., $r = DL + k\sigma\sqrt{L}$, where k is the safety stock factor.
- y $(0 \le y < 1)$ is the percentage of defective items produced in each lot-size Q.
- The length of vendor's production cycle is $nQ(1-y)/D$, and the length of buyer's ordering cycle is $Q(1-y)/D$.
- The screening rate x is fixed and is greater than the demand rate, i.e., $x > D$.
- The vendor pays a warranty cost for each defective item produced.
- The marginal profit per unit is same as the cost of lost demand (opportunity cost) per unit.
- Investment is made by the vendor in order to improve the production process quality in terms of buying new equipment, worker training, improving machine maintenance and repair, etc. The logarithmic investment function $I(y)$ given by Porteus in [22] is assumed as $I(y) = \frac{1}{\delta}\ln\left(\frac{y_0}{y}\right)$, where δ is the percentage decrease in y for a dollar increase in investment and y_0 is the percentage of defective items produced prior to investment.

2.2 *Buyer's Perspective*

Assuming the lead-time demand X to be a random variable with associated pdf $f_x(X)$, the expected shortage at the end of a cycle is given by

$$E(X-r)^+ = \int_r^\infty (x-r)f_x(X)dx \tag{1}$$

Since β is assumed to be the percentage of backorders, therefore the expected number of backorders per cycle is $\beta E(X-r)^+$. Thus, the expected stockout cost per unit time is

$$\frac{D}{Q(1-y)}\Big[\pi_x\beta + \pi_0(1-\beta)\Big]E(X-r)^+ \tag{2}$$

Further, the backorder ratio, β, is taken to be a control parameter and is assumed to be in proportion to the unit price discount offered by the buyer. That is, $\beta = (\beta_0\pi_x)/\pi_0$, where $0 \le \beta_0 < 1$ and $0 \le \pi_x \le \pi_0$.

Therefore, the backorder price discount offered by the buyer for each unit of the item, π_x, may be treated as a decision variable in place of the backorder rate, β. Hence, the expected stockout cost per unit time may be written as

$$\frac{D}{Q(1-y)}\left[(\beta_0\pi_x^2/\pi_0)+\pi_0-\beta_0\pi_x\right]E(X-r)^+ \tag{3}$$

Also, the expected net inventory level for non-defective items (just before the arrival of an order) is

$$r-DL+(1-\beta)E(X-r)^+ \tag{4}$$

And, the expected net inventory level for non-defective items (just after the shipment) is

$$Q(1-y)+r-DL+(1-\beta)E(X-r)^+ \tag{5}$$

Thus, the average non-defective inventory per cycle is given by:

$$\frac{Q(1-y)}{2}+r-DL+(1-\beta)E(X-r)^+ \tag{6}$$

Further, on the arrival of the order, the buyer inspects the items at a fixed screening rate x. Here, as is suggested by Dey and Giri in [4], it is assumed that the screening process is nondestructive and error-free and that the buyer incurs two types of holding costs—for defective items and non-defective items (including those defective items which have not yet been detected before the end of the screening time). So, the average inventory for non-defective items is

$$\frac{nQ(1-y)}{D}\left[k\sigma\sqrt{L}+\frac{Q(1-y)}{2}+\frac{DQy}{2x(1-y)}+(1-\beta)E(X-r)^+\right] \tag{7}$$

where

$$r=DL+k\sigma\sqrt{L} \tag{8}$$

Similarly, the average inventory level for defective items is given by [4]

$$nQ^2y[\frac{(1-y)}{D}-\frac{1}{2x}] \tag{9}$$

The resulting expected annual total cost for the buyer, including the shipment cost, ordering cost, inventory holding cost, shortage cost, and screening cost is, therefore, given by

$$\begin{aligned} ETCB(Q,k,n,y,\pi_x) &= \frac{D(A+nF)}{nQ(1-y)} + h_{b1}\left(Qy - \frac{DQy}{2x(1-y)}\right) \\ &+ h_{b2}\left[k\sigma\sqrt{L} + \frac{Q(1-y)}{2} + (1 - \frac{\beta_0\pi_x}{\pi_0})E(X-r)^+ + \frac{DQy}{2x(1-y)}\right] \\ &+ \frac{D}{Q(1-y)}\left[\frac{\beta_0\pi_x^2}{\pi_0} + \pi_0 - \beta_0\pi_x\right]E(X-r)^+ + \frac{sD}{1-y} \end{aligned} \tag{10}$$

where $0 \leq \pi_x \leq \pi_0$, $0 < y \leq y_0$, and $E(X-r)^+$ is the expected demand shortage at the end of the cycle. Since $r = DL + k\sigma\sqrt{L}$, then the expected demand shortage at the end of the cycle is obtained as

$$E(X-r)^+ = \sigma\sqrt{L}\psi(k) \tag{11}$$

where $\psi(k) \equiv \phi(k) - k[1-\phi(k)] > 0$.

Here, $\psi(k)$ and $\phi(k)$ denote the standard normal probability density function and distribution function, respectively. Putting these values in (11), the cost expression for the buyer is obtained as

$$\begin{aligned} ETCB(Q,k,n,y,\pi_x) &= \frac{D(A+nF)}{nQ(1-y)} + h_{b1}\left(Qy - \frac{DQy}{2x(1-y)}\right) + \frac{sD}{1-y} \\ &+ h_{b2}\left[k\sigma\sqrt{L} + \frac{Q(1-y)}{2} + (1 - \frac{\beta_0\pi_x}{\pi_0})\sigma\sqrt{L}\psi(k) + \frac{DQy}{2x(1-y)}\right] + \\ &+ \frac{D}{Q(1-y)}\left[\pi_0 - \beta_0\pi_x + \frac{\beta_0\pi_x^2}{\pi_0}\right]\sigma\sqrt{L}\psi(k) \end{aligned} \tag{12}$$

where $0 < y \leq y_0$, $0 \leq \pi_x \leq \pi_0$.

Since $\beta = \frac{\beta_0\pi_x}{\pi_0}$, thus, when $\beta = \beta_0$, we get $\pi_x = \pi_0$ and

$$\pi_0 - \beta_0\pi_x + \frac{\beta_0\pi_x^2}{\pi_0} = \pi_0 = \pi_x \tag{13}$$

Thus, when $\beta = \beta_0$, we get the case where no price discount is offered, then Eq. (12) reduces to

$$\begin{aligned} ETCB(Q,k,n,y,\pi_x) &= \frac{D(A+nF)}{nQ(1-y)} + h_{b1}(Qy - \frac{DQy}{2x(1-y)}) + \frac{sD}{1-y} \\ &+ h_{b2}[k\sigma\sqrt{L} + \frac{Q(1-y)}{2} + (1-\beta_0)\sigma\sqrt{L}\psi(k) + \frac{DQy}{2x(1-y)}] \\ &+ \frac{D}{Q(1-y)}\pi_x\sigma\sqrt{L}\psi(k) \end{aligned} \tag{14}$$

Further, if $\beta = \beta_0 = 1$ (fully backordered), then $\pi_x = \pi_0$ and the buyer's cost expression reduces to,

$$
\begin{aligned}
ETCB(Q,k,n,y,\pi_x) = {} & \frac{D(A+nF)}{nQ(1-y)} + h_{b1}(Qy - \frac{DQy}{2x(1-y)}) \\
& + h_{b2}[k\sigma\sqrt{L} + \frac{Q(1-y)}{2} + \frac{DQy}{2x(1-y)}] \\
& + \frac{D}{Q(1-y)}\pi_x\sigma\sqrt{L}\psi(k) + \frac{sD}{1-y}
\end{aligned} \tag{15}
$$

In this case, π_x is no longer the marginal price discount but the actual shortage cost is per unit item, which is the same as obtained by Dey and Giri in [4].

2.3 Vendor's Perspective

The vendor produces Q items in the first instance and delivers them to the buyer. After that, the vendor delivers a quantity of Q units to the buyer every T units of time where $T = \frac{Q(1-y)}{D}$. This process continues till the completion of the vendor's production run.

The average inventory holding area for the vendor is as given by Huang in [13]

$$
ETCV(Q,n,y) = h_v\frac{Q}{2}\left[(n-1)-(n-2)\frac{Dp}{1-y}\right] \tag{16}
$$

The annual expected total cost obtained by the vendor is the sum of the setup cost, holding cost, and warranty cost for the defective items (as given by Huang [13]).

$$
ETCV(Q,n,y) = \frac{BD}{nQ(1-y)} + h_v\frac{Q}{2}\left[(n-1)-(n-2)\frac{Dp}{1-y}\right] + \frac{wDy}{1-y} \tag{17}
$$

Adding the investment cut for improvement of production process quality as given by Dey and Giri [4], the expected annual total cost of the vendor can be derived as

$$
\begin{aligned}
ETCV(Q,k,y,n,\pi_x) = {} & \frac{BD}{nQ(1-y)} + h_v\frac{Q}{2}\left[(n-1)-(n-2)\frac{Dp}{1-y}\right] \\
& + \frac{wDy}{1-y} + \frac{\eta}{\delta}\ln\left(\frac{y_0}{y}\right)
\end{aligned} \tag{18}
$$

where η is the fractional opportunity cost.

2.4 Integrated Approach

The expected annual cost of the integrated system is the sum of the vendor's and the buyer's expected annual costs and is therefore obtained as

$$\begin{aligned} ETC(Q, y, k, n, \pi_x) = {} & \frac{D(A+B+nF)}{nQ(1-y)} + h_{b1}\left[Qy - \frac{DQy}{2x(1-y)}\right] + \frac{D(s+wy)}{1-y} \\ & + h_{b2}\left[k\sigma\sqrt{L} + \frac{Q(1-y)}{2} + \frac{DQy}{2x(1-y)} + \left(1 - \frac{\beta_0\pi_x}{\pi_0}\right)\sigma\sqrt{L}\psi(k)\right] \\ & + \frac{\eta}{\delta}\ln\left(\frac{y_0}{y}\right) + \frac{D}{Q(1-y)}G1(\pi_x)\sigma\sqrt{L}\psi(k) \\ & + h_v\frac{Q}{2}\left[(n-1) - (n-2)\frac{Dp}{1-y}\right] \end{aligned} \tag{19}$$

where

$$G1(\pi_x) = (\pi_0 - \beta_0\pi_x + \beta_0\pi_x^2/\pi_0) < 0 \tag{20}$$

(because $\frac{\pi_0}{\pi_x} > \beta_0\left(1 - \frac{\pi_x}{\pi_0}\right) > 0$)

3 Solution Procedure

In the above objective function, the control parameters are Q, y, k, n, and π_x. It is very difficult to show analytically that ETC is a convex function in all the decision variables. However, it can be verified numerically that for given values of n (positive integer), $y (0 < y \leq y_0 < 1)$, and π_x, the total cost function ETC is convex in Q and k (refer Fig. 4). It can also be shown that, all other parameters being known, ETC is convex w.r.t. y (refer Fig. 3). In the following section, a solution procedure is outlined to derive the optimal values of Q, y, k, n, and π_x such that the joint expected annual total cost ETC is minimized.

The total cost function ETC can be shown to be convex in n and π_x, respectively, as

$$\frac{\partial^2 ETC}{\partial n^2} = \frac{2D(A+B)}{n^3 Q(1-y)} > 0 \;\forall\, n \geq 1 \tag{21}$$

$$\frac{\partial^2 ETC}{\partial \pi_x^2} = \frac{2D\sqrt{L}\sigma\beta_0\psi(k)}{Q(1-y)\pi_0} > 0 \tag{22}$$

Keeping in mind the potential non-convexity of the model, an iterative procedure is suggested to derive the optimal values of the control parameters as follows:

For fixed n, the first derivative of ETC with respect to k is put to zero. That is,

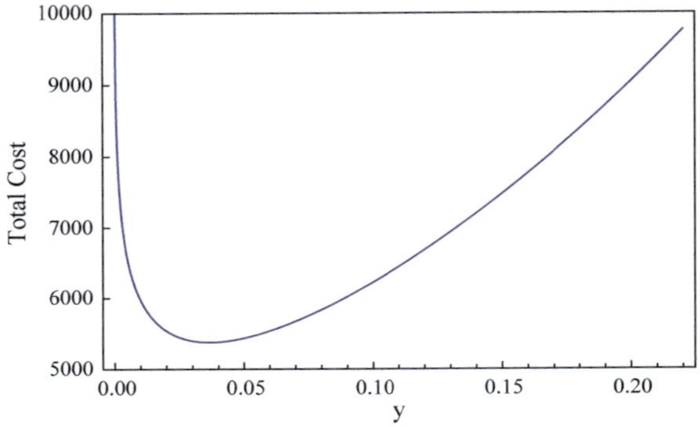

Fig. 3 ETC w.r.ty

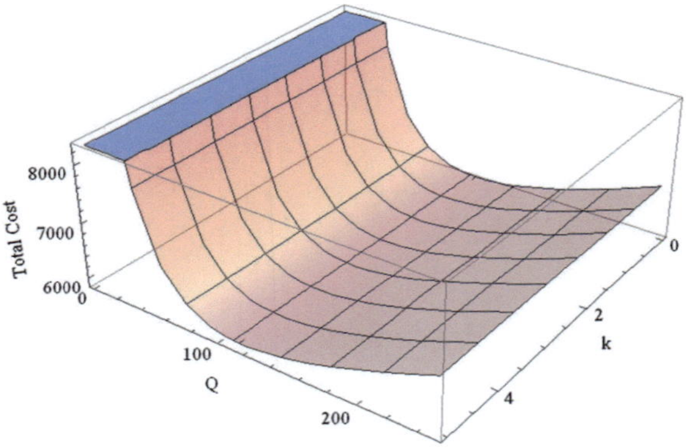

Fig. 4 ETC w.r.t Q, k

$$\frac{\partial ETC}{\partial k} = h_{b2}\left[\sigma\sqrt{L} + \left(1 - \frac{\beta_0\pi_x}{\pi_0}\right)\sigma\sqrt{L}\psi'(k)\right] + \left[\frac{DG1(\pi_x)\sigma\sqrt{L}\psi'(k)}{Q(1-y)}\right] = 0 \tag{23}$$

Or,

$$\frac{\partial ETC}{\partial k} = h_{b2}\left[\sigma\sqrt{L} + \left(1 - \frac{\beta_0\pi_x}{\pi_0}\right)\sigma\sqrt{L}(F(k) - 1)\right] + \frac{DG1(\pi_x)\sigma\sqrt{L}(F(k) - 1)}{Q(1-y)} = 0 \tag{24}$$

where $F(\cdot)$ is the standard normal distribution function.

This implies

$$\overline{F(k)} = \frac{h_{b2}}{H1} = H2(say) \tag{25}$$

where $H1 = h_{b2}(1 - \beta_0\pi_x/\pi_0) + \frac{DG1(\pi_x)}{Q(1-y)}$ and $\overline{F}(\cdot) = 1 - F(\cdot)$.

Similarly we have

$$\frac{\partial ETC}{\partial Q} = -\frac{DG(n)}{Q^2(1-y)} + yh_{b1}\left(1 - \frac{D}{2x(1-y)}\right) - \frac{D\sqrt{(}L)G1(\pi_x)\sigma\psi(k)}{Q^2(1-y)}$$
$$+ h_{b2}\left[\frac{1-y}{2} + \frac{Dy}{2x(1-y)}\right] + \frac{h_v}{2}\left[(n-1) - (n-1)\frac{Dp}{1-y}\right] = 0 \tag{26}$$

$$\frac{\partial ETC}{\partial y} = \frac{Dw}{1-y} + \frac{D(t+wy)}{(1-y)^2} - \frac{\eta}{y\delta} + \left[\frac{D\sqrt{(}L)G1(\pi_x)\sigma\psi(k)}{Q(1-y)^2}\right]$$
$$+ \frac{DG(n)}{Q(1-y)^2} + Qh_{b1}\left[1 - \frac{D}{2x(1-y)} - \frac{Dy}{2x(1-y)^2}\right]$$
$$- h_{b2}[-\frac{Q}{2} + \frac{DQ}{2x(1-y)} + \frac{DQy}{2x(1-y)^2}] + \frac{Qh_v}{2}\left[\frac{D(n-2)p}{(1-y)^2}\right] = 0 \tag{27}$$

$$\frac{\partial ETC}{\partial \pi_x} = \frac{D\sqrt{(}L)(-\beta_0 + 2\beta_0\pi_x/\pi_0)\sigma\sqrt{(}L)\psi(k)}{Q(1-y)} - \left[\frac{\beta_0\sqrt{(}L)\sigma h_{b2}\psi(k)}{\pi_0}\right] = 0 \tag{28}$$

On simplification, (26) reduces to:

$$Q = \sqrt{\frac{DG(n) + G1(\pi_x)D\sigma\sqrt{(}L)\psi(k)}{H(n, y)}} \tag{29}$$

where

$$H(n, y) = h_{b1}[y(1-y) - \frac{Dy}{2x}] + h_{b2}[\frac{(1-y)^2}{2} + \frac{Dy}{2x}] + h_v/2 - (n-2)Dp + (n-1)(1-y) \tag{30}$$

From (28), we get

$$\pi_x = \frac{D\pi_0 + Qh_{b2}(1-y)}{2D} \tag{31}$$

It is obvious that the control parameters are not independent of each other. So, to obtain a solution, we adapt the iterative algorithm proposed by Ben-Daya and Hariga in [3] and adapted by Dey and Giri in [4]. First, the algorithm is initiated by setting $y = y_0$, where y_0 is the original percentage of defective items produced. Next, an initial value of Q is calculated by setting the stochastic parameter of Q equal to zero in (29). After calculating Q, the value of k and π_x is obtained using (25) and (31), respectively. Then, using these values, the value of y is updated. This process continues till the minimum cost is determined for the given value of n. The value of n is then updated, and the whole process is repeated till the minimum cost is obtained. The values of the control parameters corresponding to the minimum cost represent the optimal solution. It is to be noted that if the updated value of y is found to be greater than the initial value y_0, then the updated value is rejected. This follows intuitively because the purpose of vendor investment is to improve the production process and failure to do so negates the purpose itself. The method is encapsulated in an algorithm below:

3.1 The Algorithm:

Step 1: Set $ETC^* = \infty$ and $n = 1$;

Step 2: Set $y = y_0$ and compute $Q_0 = \sqrt{DG(n)/H(n, y)}$;

Step 3: Find k from $\psi(k) = \int_k^{\infty}(z - k)\phi(z)dz$ from (25).

Step 4: Compute π_x using (31).

Step 5: Compute y from (27).

Step 6: Compute Q, from (29) using k, y and π_x. If $|Q - Q_0| \leq \epsilon$, compute $ETC(Q, k, y, n, \pi_x)$ and go to step 7. Else set $Q_0 = Q$ and go to step 3;

Step 7: If $ETC^* \geq ETC$, Set $ETC^* = ETC$, $Q^* = Q$, $y^* = y$, $k^* = k$, $n = n + 1$ and go to step 2; Else $n^* = n - 1$ and stop.

The corresponding values of the control parameter for $n^* = n - 1$ give the optimal solution.

Now, partially deriving ETC with respect to w, δ, and y_0 we get

$$\frac{\partial ETC}{\partial w} = \frac{Dy}{1 - y} > 0 \tag{32}$$

$$\frac{\partial ETC}{\partial \delta} = -\frac{\eta}{\delta^2} \ln\left(\frac{y_0}{y}\right) < 0 \tag{33}$$

$$\frac{\partial ETC}{\partial y_0} = \frac{\eta}{y_0 \delta} > 0 \tag{34}$$

From above Eqs. (32)–(34), we observe that ETC increases with an increase in warranty cost w and also with an increase in the original percentage of defective items y_0. For a higher warranty cost, the integrated systems expected total cost will increase and there will also be an increase in the backorder price discount. Also, production of items of very poor quality will result in an increase in the expected total cost of the system. Increasing δ implies that the number of defective items decreases with an increase in the investment amount and, thus, the expected total cost of the system. That is, it costs less to improve the production process quality of the system. To further highlight the effects of process quality, the investment, and other model parameters on the optimal inventory decisions, numerical studies are presented in the following section.

4 Numerical Examples

For numerical studies, we consider the following data set: $D = 1000$, $P = 3200$, $A = 50$, $B = 400$, $F = 35$, $L = 10/365$, $h_v = 4$, $h_{b1} = 6$, $h_{b2} = 10$, $t = 0.25$, $s = 100$, $x = 2152$, $w = 24$, $y = 0.22$, $\eta = 0.2$, $\delta = 0.0002$, $\sigma = 5$, $\pi_0 = 150$, $\beta_0 = 0.2$.

4.1 *With Constant Lead-Time*

Tables 1, 2, 3, and 4 show the effects of various parameters on the integrated system when the lead-time is constant. (Here z^* denotes π_x^* which is the optimal value of the value of the unit backorder price discount.)

4.2 *With Variable Lead-Time*

Tables 5, 6, 7, and 8 show the effects of various parameters on the integrated system when the lead-time is taken to be dependent on the lot-size in the form $L = pQ + b$.

Tables 1 and 5 show that the production lot-size increases with an increase in demand rate, which is very obvious since the buyer would need to place an order of a larger quantity to satisfy the increase in demand. Also, there is an increase in the quantity of both defective and non-defective items produced with an increase in

Table 1 Effect of demand rate

D	Q^*	n^*	y^*	ETC^*	$I(y^*)$	k^*	π_x^*
800	84.3442	6	0.044	4911.59	1593.71	2.40639	75.5036
900	90.1529	6	0.040	5153.64	1698.4	2.38026	75.4807
1000	95.8221	6	0.036	5375.09	1792.89	2.35630	75.4616
1100	91.6176	7	0.033	5576.63	1877.08	2.37178	75.4024
1200	96.7118	7	0.031	5762.07	1956.02	2.35073	75.3904

Table 2 Effect of production rate

P	Q^*	n^*	y^*	ETC^*	$I(y^*)$	k^*	π_x^*
2800	87.9689	7	0.036	5335.41	1788.90	2.38794	75.4327
3000	87.1628	7	0.036	5357.04	1790.10	2.39131	75.4198
3200	95.8221	6	0.036	5375.09	1792.89	2.35630	75.4616
3400	95.2388	6	0.036	5389.66	1793.67	2.35855	75.4588
3600	94.7290	6	0.036	5402.55	1794.37	2.36053	75.4563

Table 3 Effect of parameters w and δ

		Q^*	n^*	y^*	ETC^*	$I(y^*)$
w	20	96.154	6	0.042	5210.09	1633.98
	24	95.8221	6	0.036	5375.09	1792.89
	30	95.4785	6	0.030	5580.52	1991.91
δ	0.00020	95.8221	6	0.036	5375.09	1792.89
	0.00024	95.5229	6	0.030	5062.51	1636.14
	0.00030	95.2203	6	0.024	4718.35	1449.84

Table 4 Effect of y_0

y_0	Q^*	n^*	y^*	ETC^*	$I(y^*)$
0.040	95.8221	6	0.036	3670.34	88.1429
0.100	95.8222	6	0.036	4586.63	1004.43
0.220	95.8221	6	0.036	5375.09	1792.89
0.418	95.8221	6	0.036	6016.94	2434.74
0.680	95.8221	6	0.036	6503.55	2921.34

Table 5 Effect of demand rate

D	Q^*	n^*	y^*	ETC^*	$I(y^*)$	k^*	π_x^*
800	75.5015	6	0.044	4915.07	1593.91	2.40789	75.5015
900	89.7808	6	0.040	5157.81	1698.60	2.38173	75.4787
1000	86.1346	7	0.036	5379.64	1791.32	2.39559	75.4149
1100	91.2576	7	0.033	5581.09	1877.27	2.37316	75.4008
1200	96.3325	7	0.031	5767.17	1956.20	2.35211	75.3889

Table 6 Effect of production rate

P	Q^*	n^*	y^*	ETC^*	$I(y^*)$	k^*	π_x^*
2800	87.5897	7	0.037	5340.82	1789.11	2.38946	75.4218
3000	86.8042	7	0.037	5361.61	1790.30	2.39276	75.4181
3200	86.1346	7	0.036	5379.64	1791.21	2.39559	75.4149
3400	85.5569	7	0.036	5395.42	1792.21	2.39804	75.4121
3600	94.3689	6	0.036	5406.08	1794.54	2.36189	75.4546

Table 7 Effect of parameters w and δ

		Q^*	n^*	y^*	ETC^*	$I(y^*)$
w	20	86.4461	7	0.043	5214.35	1632.12
	24	86.1346	7	0.036	5379.64	1791.32
	30	95.0873	6	0.030	5585.32	1992.07
δ	0.00020	86.1346	7	0.036	5379.64	1791.32
	0.00024	95.1318	6	0.030	5067.32	1636.43
	0.00030	94.8299	6	0.024	4723.12	1449.87

Table 8 Effect of y_0

y_0	Q^*	n^*	y^*	ETC^*	$I(y^*)$
0.040	86.1346	7	0.037	3674.90	86.5660
0.100	86.1346	7	0.037	4591.19	1002.86
0.220	86.1346	7	0.036	5379.64	1792.32
0.418	86.1345	7	0.036	6021.50	2433.18
0.680	86.1344	7	0.036	6508.11	2919.78

the lot-size, and consequently, the amount of investment required to optimize the expected annual total cost also increases. Again, we see that an increase in demand causes a decrease in the safety stock factor and the unit backorder price discount also decreases. All these intuitively correct effects are illustrated numerically.

Tables 2 and 6 show that an increase in the production rate simultaneously increases the number of defective and non-defective items produced and hence results in an increase in the amount of investment required to improve the production process quality which in turn increases the expected total cost of the system.

A significant conclusion that can be reached from Tables 4 and 8 is that the original quality and the investment required to improve the production process quality are not independent. That is, the original production process quality decides the necessity of an investment and the extent of it being beneficial to the supply chain. This is evident from Table 1 which clearly shows that investment to improve the production process quality is not needed when the original percentage of defectives is very low. However, the amount of investment required to optimize the integrated system increases noticeably with an increase in the value of y_0.

Tables 3 and 7 depict that the optimal total cost incurred by the supply chain increases with an increase in the warranty cost w paid by the vendor. Also, an increase in warranty cost results in a decrease in the optimal value of the percentage of defective items. This is intuitively correct because, as the vendor is required to pay a higher warranty cost as a penalty for defective item being produced, a considerable decrease in the number of defective items produced would reasonably be favorable for him. Following the same logic, an increase in the value of δ should imply a decrease in the expected annual total cost of the system as also shown in Tables 3 and 7.

5 Managerial Insights and Some Concluding Remarks

The paper presents a single-vendor single-buyer integrated imperfect production–inventory model with vendor investment and backorder price discount. It is assumed that the vendor makes an investment for improving the production process quality. A simple iterative procedure is suggested to obtain an optimal solution of the proposed model in terms of minimizing the total cost incurred. It is shown through numerical examples, under various scenarios, that the production rate of defective items is reduced when the vendor makes an investment. This investment made by the vendor helps in reducing the expected annual total cost, thereby resulting in a better optimized integrated system. It is also evident from the numerical studies that an increased investment is required for an increased demand rate to minimize the total cost. As a scope of future research work, the proposed model can be extended in terms of variable shipment size, inspection errors, controllable lead-time, multiple buyers, etc.

References

1. Banerjee, A.: A joint economic-lot-size model for purchaser and vendor. Decis. Sci. **17**, 292–311 (1986)
2. Ben-Daya, M., Hariga, M.: Economic lot scheduling problem with imperfect production process. J. Oper. Res. Soc. **51**, 875–881 (2000)
3. Ben-Daya, M., Hariga, M.: Integrated single vendor single buyer model with stochastic demand and variable lead-time. Int. J. Prod. Econ. **92**, 75–80 (2004)
4. Dey, O., Giri, B.C.: Optimal vendor investment for reducing defect rate in a vendor?buyer integrated system with imperfect production process. Int. J. Prod. Econ. **155**, 222–228 (2014)
5. Glock, C.H.: A comment: integrated single vendor-single buyer model with stochastic demand and variable lead time. Int. J. Prod. Econ. **122**, 790–792 (2009)
6. Glock, C.H.: Lead timer eduction strategies in a single-vendorsingle-buyer integrated inventory model with lot size-dependent lead times and stochastic demand. Int. J. Prod. Econ. **136**, 37–44 (2012)
7. Goyal, S.K.: An integrated inventory model for a single supplier-single customer problem. Int. J. Prod. Res. **15**, 107–111 (1976)
8. Goyal, S.K.: A joint economic-lot-size model for purchaser and vendor: a comment. Decis. Sci. **19**, 236–241 (1988)
9. Goyal-Cardenas-Barron, L.E.: The economic production quantity (EPQ) with shortage derived algebraically. Int. J. Prod. Econ. **70**, 289–292 (2002)
10. Ha, D., Kim, S.L.: Implementation of JIT purchasing: an integrated approach. Prod. Plan. Control. **8**, 152–157 (1997)

11. Hill, R.M.: The optimal production and shipment policy for the single-vendor single-buyer integrated production-inventory problem. Int. J. Prod. Res. **37**, 2463–2475 (1999)
12. Hsiao, Y.-C.: Integrated logistic and inventory model for a two-stage supply chain controlled by the reorder and shipping points with sharing information. Int. J. Prod. Econ. **115**, 229–235 (2008)
13. Huang, C.K.: An optimal policy for a single-vendor single-buyer integrated production-inventory problem with process unreliability consideration. Int. J. Prod. Econ. **91**, 91–98 (2004)
14. Jaggi, C.K., Arneja, N.: Periodic inventory model with unstable lead-time and set-up cost with backorder price discount. Int J. Appl. Decis. Sci. **3**(1), 53–73 (2010)
15. Jauhari, W.A.: Lot size decisions for vendor-buyer system with quantity discount, partial back-order and stochastic demand. Adv. Oper. Res. Article ID-597626, 7p (2014)
16. Jindal, P.A.S.: Integrated supply chain inventory model with quality improvement involving controllable lead-time and Backorder price discount. Int. J. Ind. Eng. Comput. **7**(3), 463–480 (2016). https://doi.org/10.5267/j.ijec.2015.12.003
17. Lin, Y.-J.: An integrated vendor-buyer inventory model with backorder price discount and effective investment to reduce ordering cost. Comput. Ind. Eng. **56**, 1597–1606 (2009)
18. Lin, H.J.: An integrated supply chain inventory model with imperfect-quality items, controllable lead time demand and distribution-free demand. Yugoslav J. Oper. Res. **23**, 87–109 (2013)
19. Ouyang, L.Y., Wu, K.S., Ho, C.H.: Analysis of optimal vendor-buyer integrated inventory policy involving defective items. Int. J. Adv. Manuf. Technol. **29**, 1232–1245 (2006)
20. Pal, M., Chandra, S.: Aperiodic review inventory model with stock dependent demand,permissible delay in payment and price discounts on backorder. Yugoslav J. Oper. Res. **24**(1), 99–110 (2012). https://doi.org/10.2298/YJOR120512017P
21. Pan, J.C.H., Yang, J.S.: A study of an integrated inventory with controllable lead time. Int. J. Prod. Res. **40**, 1263–1273 (2002)
22. Porteus, E.L.: Optimal lot sizing, process quality improvement and setup cost reduction. Oper. Res. **36**, 137–144 (1986)
23. Rosenblatt, M.J., Lee, H.L.: Economic production cycles with imperfect production process. IIE Trans. **18**, 48–55 (1986)
24. Yang, J.S., Pan, C.H.J.: Just-in-time purchasing: an integrated inventory model involving deterministic variable lead time and quality improvement investment. Int. J. Prod. Res. **42**(5), 853–863 (2004)
25. Zhang, X., Gerchak, Y.: Joint lot sizing and inspection policy in an EOQ model with random yield. IIE Trans. **22**, 41–47 (1990)

Further Reading

26. Jana, D.K., Chakraborty, D., Maiti, M.: A random fuzzy production inventory problem with backorder rate based on controllable preparation time and safety factor via genetic algorithm
27. Jana, D.K., Maity, K., Roy, T.K.: Multi-objective imperfect production inventory model in fuzzy rough environment via genetic algorithm

Fuzzy Time Series Model for Unequal Interval Length Using Genetic Algorithm

Shanoli Samui Pal and Samarjit Kar

Abstract Time series analysis and forecasting depend on the observed historical data. As it deals with numerous number of information, imprecision occurs due to many reasons like truncation error, mechanical fault, noise. In such cases, fuzzy concept works well to handle impreciseness. Fuzzy time series forecasting methodology consists of defining universe of discourse (UOD), fuzzification of time series data points, assigning relationships between consecutive data points and defuzzification to get back the forecasting results in real domain. In this paper, UOD of a time series has been defined, unequal partitions of the UOD are established and then relationship among consecutive data points is evaluated to get the forecasting models. Genetic algorithm has been used in partitioning the UOD unequally and establishing the relationship. Forecasting model is applied on enrollment of University of Alabama, BSE sensex time series, and Shenzhen stock exchange data. The results are compared with conventional fuzzy time series models.

Keywords Fuzzy time series · Genetic algorithm · Time series

1 Introduction

Time series analysis and forecasting are used for predicting the future pattern based on historical information. There are several existing works which are about new forecasting models.Uncertainty handling in time series forecasting has been achieved by using the fuzzy logic. Uncertainty can be emerged due to several reasons like mechanical noise, truncation error, missing information. Fuzzy time series forecasting model is such an attempt in which fuzzy logic has been first introduced by Song and Chissom [13, 14] in their time-variant and time-invariant models. There are four

S. S. Pal (✉) · S. Kar
National Institute of Technology, Durgapur 713209, WB, India
e-mail: shanoli.pal@gmail.com

S. Kar
e-mail: dr.samarjitkar@gmail.com

P. Chandra et al. (eds.), *Information Technology and Applied Mathematics*,
Advances in Intelligent Systems and Computing 699,
https://doi.org/10.1007/978-981-10-7590-2_15

basic steps in these conventional fuzzy time series forecasting models [13, 14]: (i) deciding the universe of discourse, length, and number of intervals, i.e., partitioning the UOD, (ii) fuzzify the time series data points by defining fuzzy sets, (iii) establishing fuzzy logical relationship (FLR) and grouping of FLR (FLRG) if any exits, and (iv) defuzzifying the obtained result to get the final output. Fuzzy time series forecasting models are studied to improve its drawbacks like computational complexity and model accuracy. Also, these forecasting models are largely used in several areas of application like stock market, enrollment data, rainfall prediction, exchange rates, tourism [15, 16].

Researchers proposed several concepts to improve one or more of these steps to improve forecasting accuracy or to reduce the computational burden. Step (i), Step (ii), and Step (iv) are mostly interesting areas in which several ideas have been proposed from the very beginning of the era of fuzzy time series. Partitioning the interval equally or unequally, weighted or unweighted in conventional step (i) has been already examined to obtain better accuracy. Assigning fuzzy membership to fuzzify the data points is also studied to improve the forecasting technique. Different defuzzifying techniques are adopted to get better results.

Some of the important improvements on fuzzy time series are briefly discussed here. Fuzzy time series together with real coded genetic algorithm was used by Chen and Chung [2] to adjust the length of intervals which resulted in unequal interval partitioning of UOD. Also, higher-order fuzzy logical relationships was adopted in the work proposed by Chen and Chung [2]. In conventional fuzzy forecasting model, the equal length intervals was considered in universe of discourse. But Chen and Chung [2] tried to adjust the length of each interval in the universe of discourse as it would affect the forecasting accuracy rate, and they showed that the model was better than the conventional models. Lee et al. [10] modified the weighted fuzzy time series forecasting. Chen [3] proposed fuzzy time series model where fuzzy logical relationships were grouped into rules to reduce the computational burden. Yu [17] developed fuzzy time series forecasting model with the weighted fuzzy logical relationships.

In most of the proposed methodologies, defuzzifying the obtained outcome and forecasting the output are almost same as Song and Chissom's model [13, 14] and Chen's model [3].

Pal et al. [12] have proposed a hybrid model to modify the defuzzification step where intervals are equally partitioned and FLRs with first-order have been established. In this work, no particular fuzzy membership function is used to fuzzify the time series data points. It finds the intervals corresponding to each time series data point in which it belongs and establishes the FLR with first-order between consecutive data points. Mid-points are adjusted at defuzzification step by approximating the original values in corresponding intervals using real coded GA. Halton sequence [5] has been used to initialize the population of GA. Neural network with back propagation learning (BPNN) was used to learn the pattern of the time series and to forecast from trained network.

The third step of establishing FLR and FLRG proposed till now is of two types which are followed from conventional fuzzy time series forecasting models. The one proposed by Song and Chissom [13, 14] has computational burden. And another proposed by Chen [3] suffers from fixed UOD, and there is no scope to enlarge the universe as per our knowledge. This work improves these limitations by finding the relationship among mid-points of consecutive data points by real coded GA, where the higher order of FLRs have been also considered. Here, unequal length intervals have been decided by real coded GA. Double GA is used here to get the forecasting model.

The remaining of this paper is organized as follows. In Sect. 2, we review the basic concepts of fuzzy time series and GA. Section 3 includes the proposed model description with algorithmic steps. Results are discussed with data description in Sect. 4. Finally, Sect. 5 gives the conclusion and future scopes. Appendix is in Sect. 6.

2 Preliminaries

In this section, fuzzy time series, FLR and FLRG, and genetic algorithm (GA) have been presented briefly.

2.1 *Fuzzy Time Series*

Fuzzy time series are defined by fuzzy set of real-time series data points. Fuzzy set and its membership function are defined as below.

(1) Fuzzy set and its membership function: Let A be a fuzzy set defined in the universe of discourse $U = \{u_1, u_2, u_3, ..., u_n\}$ as follows:

$$A = f_A(u_1)/u_1 + f_A(u_2)/u_2 + ... + f_A(u_n)/u_n, \tag{1}$$

where f_A denotes the membership of the fuzzy set A, $f_A : U \rightarrow [0, 1]$, $f_A(u_i) \in [0, 1]$, $1 \leq i \leq n$, and "+" denotes the union operation.

Let $X(t)$, $(t = \ ...0,\ 1,\ 2,\ ...)$ be the UOD defined in real domain. If $f_i(t)$, $(t = \ ...0,\ 1,\ 2,\ ...)$ can be defined in UOD $X(t)$ and $F(t)$ is a collection of such $f_i(t)$s like $f_1(t),\ f_2(t),\ f_3(t),\ ...$ then $F(t)$ is called a fuzzy time series defined on UOD $X(t)$.

(2) Fuzzy relationship: Fuzzy relationship between consecutive fuzzy data points $f(t)$ and $f(t-1)$ is denoted by $\check{R}(t,\ t-1)$ and defined by $f(t) = f(t-1)o\check{R}(t, t-1)$, where "$o$" is a composition operator. Establishing FLR can be explained as $f(t)$ is caused by $f(t-1)$ with the help of $\check{R}$, and it is denoted as $f(t-1) \rightarrow f(t)$, where both $f(t)$ and $f(t-1)$ are fuzzy sets. This relation $\check{R}$ is called first-order fuzzy logical relationship model of $F(t)$. Similarly, if $f(t)$ is caused by historical n-terms, i.e., $f(t-1)$, $f(t-2)$, ..., and $f(t-n)$, then the fuzzy logical relationship can be denoted as follows:

$$f(t-n), ..., f(t-2), f(t-1) \rightarrow f(t) \quad (2)$$

It is called the nth-order FLR model, where $n \geq 2$.

Let $f(t-1) = A_i$ and $f(t) = A_j$. The first-order FLR between two consecutive data points, $f(t-1)$ and $f(t)$, is denoted by $A_i \rightarrow A_j$.

The relation Ř was generated by matrix operation in [13, 14] with huge computational burden. To avoid this computational complexity, Chen [3] proposed an idea to get the defuzzified value which was the mid-point of the interval respective to A_j. In case of fuzzy logical relationships in group, defuzzified value would be average of the middle-points of corresponding intervals; i.e., if relationships are like following,

$$\begin{aligned} &A_{i_1}, A_{i_2}, ..., A_{i_k} \rightarrow A_{j_1} \\ &A_{i_1}, A_{i_2}, ..., A_{i_k} \rightarrow A_{j_2} \\ &. \\ &. \\ &. \\ &A_{i_1}, A_{i_2}, ..., A_{i_k} \rightarrow A_{j_k} \end{aligned} \quad (3)$$

then defuzzified value would be average of middle-points corresponding to A_{j_1}, A_{j_2}, ..., A_{j_k}. So, it searches the middle-points of the current value $F(t)$s, which must be inside the UOD; i.e., it always gives defuzzified value inside defined interval of UOD. Hence, there was no scope to enlarge the UOD.

In this paper, we are searching the relation Ř in the form of following expression

$$f(t+r) = m(t) * x_1 + m(t+1) * x_2 + ... + m(t+\overline{r-1}) * x_r, \quad (4)$$

where $f(t+r)$ is the current time value that depends on tth, (t+1)th, ..., $(t+\overline{r-1})$th time points. So the middle-points $m(t)$, $m(t+1)$, ..., $m(t+\overline{r-1})$ corresponding to tth, (t+1)th,..., $(t+\overline{r-1})$th time points are used here. r denotes the order of the FLR. $x_1, x_2, ..., x_r$ are parameters, whose optimum values are searched by using GA.

2.2 Genetic Algorithm (GA)

GA is a population-based optimization tool hugely studied and used in searching optimum solutions of problems. It mimics the natural evolution of genes and gives the optimum solution like survival of the most efficient. GA searches for the best possible solution which is the chromosome in GA terminology from population-based search space. Population is the pool of chromosomes. Chromosomes are combination of genes which are the parameters for the specific problem to be decided. The search procedure is repeated for few generations to get best output like our real-world phenomenon where genes are evaluated generation to generation.

Accuracy of a time series forecasting model is measured by using several error metrics like mean square error, root mean square error, absolute error. Our objective in time series forecasting is developing such model which reduces the difference between original and forecasted values to achieve high accuracy. Here, those error metrics can be considered as fitness function in GA. Chromosomes are ranked according to the fitness. Most fitted chromosomes will survive in next generation. Chromosomes with lower fitness will not survive in the following generations. The next generation consists of the chromosomes with higher fitness of previous generation, modified chromosomes after applying crossover and mutation operator, and chromosomes randomly generated by the systems. Each generation consists of same number of chromosomes and evolves toward accuracy.

Three basic operations in GA are described as follows.

1. *Selection*:
   This operator in GA is used to select the fittest chromosomes, which will carry on in next generation. Some chromosomes are selected as parent chromosomes which are used to reproduce child chromosomes for the following generation. Using this mechanism, mostly fitted chromosomes are selected more than the weakly fitted chromosomes.
2. *Crossover*:
   Objective of this operator is to reproduce child chromosomes to explore the search space. Two parent chromosomes will exchange their genes to share information. There are several mechanism of exchanging genes like single point crossover, double point crossover.
3. *Mutation*:
   This operator gives the advantage of sudden changes in genes to exploit the search space. Some genes of selected chromosomes are mutated following the probability of mutation. In general, genes are mutated randomly.

Finally, reproduced chromosomes and mutated chromosomes will continue in next generation.

3 Proposed Approach Using Double GA on Fuzzy Time Series

In this paper, GA has been used to search the optimum value of $x_1,\ x_2, \ldots,\ x_r$ by solving Eq. (4) to improve the forecasting results. For example, the optimum value of $x_1,\ x_2$ has been searched for second-order FLR like Eq. (5),

$$f(t+2) = m(t) * x_1 + m(t+1) * x_2, \tag{5}$$

and the optimum value of $x_1,\ x_2,\ x_3$ have been searched for third-order FLR like Eq. (6) using GA

$$f(t+3) = m(t) * x_1 + m(t+1) * x_2 + m(t+2) * x_3. \qquad (6)$$

Proposed approach is discussed in following algorithmic steps.

Step 1 Decide the UOD. In case enrollment data of University of Alabama, minimum and maximum values are respectively 5,127 and 27,052. So, UOD is $U =$ [5,100 27,100].

Step 2: Unequal partition of UOD
Real coded GA is used to get the unequal intervals. As double GA is used here, we denote the first one as GA-I and the next one as GA-II. Initial population with chromosome number p is generated with g number of uniformly random selected genes, whose last two points are lower and upper bound, i.e., 5,100 and 27,100, respectively. Selected genes may not be in order, so they are arranged in ascending order after selecting randomly.

Step 3: Find out the mid-points of each interval in which actual data point lies. It is possible that there exists intervals having more than one data points as well as intervals having no data points. Then, higher-order fuzzy logical relationships between consecutive data points are established. For each chromosome in the pool, follow the following steps.

Step 4: Apply GA-II with n data points and their corresponding mid-points. Higher-order fuzzy logical relationship is used here. Basically, we have tried to solve equations like Eq. (4) where $x_1, x_2, \ldots, x_r$ are optimum value, searched by GA-II.

a. Initialization of the population: Chromosomes are selected randomly from uniform distribution to initialize the population.
b. Fitness function: After initialization, current time point $f(t+r)$ has been calculated, where $t \leq n$. Here, the fitness function is the MSE calculated from the original data point. For all chromosomes in the pool, MSEs are calculated. Rank the chromosomes, and select the chromosome with minimum MSE which is the most fitted chromosome for that population. Few best chromosomes have been passed to the next generation. Crossover and mutation are operated on remaining chromosome.
c. Crossover: It randomly selects few chromosomes from remaining pool, of which two randomly selected chromosomes say, $c1$, $c2$, exchange their genes at crossover point, which is also randomly selected.
d. Mutation: Randomly selected chromosomes are modified adding random number, which is uniform(0, 1), with the values of $x_1, x_2, \ldots, x_r$, respectively.
e. Repeat some fixed number of generation from Step b to Step d, which is considered here as termination criterion. So, we have now optimum value of $x_1, x_2, \ldots, x_r$ and best calculated MSE using those $x_1, x_2, \ldots, x_r$ for each occurrence of the $(t+r)$th, $t \leq n$ data point.

Repeat **Step 4** for all chromosomes in the initial population of GA-I.

Step 5: In such way, best MSEs are assigned to each chromosome in the pool of GA-I in first generation. Best chromosomes are ranked according to the value of the fitness function. Few best chromosomes are kept and passed to the next generation to fulfill the elitism mechanism. Few worst chromosomes are deleted permanently by replacing them with random real numbers. Crossover and mutation are applied on rest of the chromosomes. Two consecutive parent chromosomes are selected, and crossover point is selected randomly. At that crossover point, two parent chromosome exchange their genes to create child chromosomes. For mutation on some chromosomes, mutation point, say mute_point, is generated randomly and gene at that point is replaced by the number selected uniformly within (mute_point −1) and (mute_point +1). All these changed chromosomes are passed to the next generation. In such way, few generations are repeated. At last, we get the best chromosome which is the unequal intervals in our case.

Mean square error (MSE) is calculated from the original and forecasted values. MSE is given in the appendix.

4 Result and Discussion

The two theories, which are adapted in this work, are (i) unequal interval lengths and (ii) artificial way of establishing relationship $\check{R}$ to decide higher-order FLRs. Effective lengths of intervals improve forecasting accuracy which is already established [7] in fuzzy time series. Next one is obvious as current situation depends on more than one previous occurrences. Performance of the proposed approach shows that these two theories work well altogether on three time series.

In this paper, proposed approach is applied on three time series data. First one is the enrollment data of the University of Alabama collected yearly for the years 1948–2009 [6, 8].

Second one is on Bombay stock exchange (BSE). BSE's popular equity index—the S&P BSE sensex- is India's mostly tracked stock market benchmark index. There are several work, i.e., stock market analysis, forecasting models on BSE data [4, 9]. In our work, BSE sensex closing price has been collected daily from January, 2014 to March, 2015 (301 data points) [1].

Another time series data on Shenzhen stock exchange has been considered here, which is one of the three stock exchanges in China. There are several existing works on Shenzhen stock exchange [11, 18]. Opening, high, low, and closing prices of 30 stocks are collected of which closing prices of one stock have been analyzed here. There are 593 data points of the first stock from April 11, 2004, to Dec 3, 2015.

Figure 1 shows the flows of the proposed method, which is also described in Sect. 3. Figure 2, Fig. 3, and Fig. 4 depict the outcomes of the proposed methodology for enrollment data, BSE sensex data, and Shenzhen stock exchange data, respectively.

Accuracy of the model is evaluated by calculating the mean square error (MSE). The forecasting performance of the proposed model is compared with the conventional fuzzy time series models [3, 6, 13, 14]. MSE values of the proposed model on enrollment data sets are given in Table 1 comparing with some conventional previous

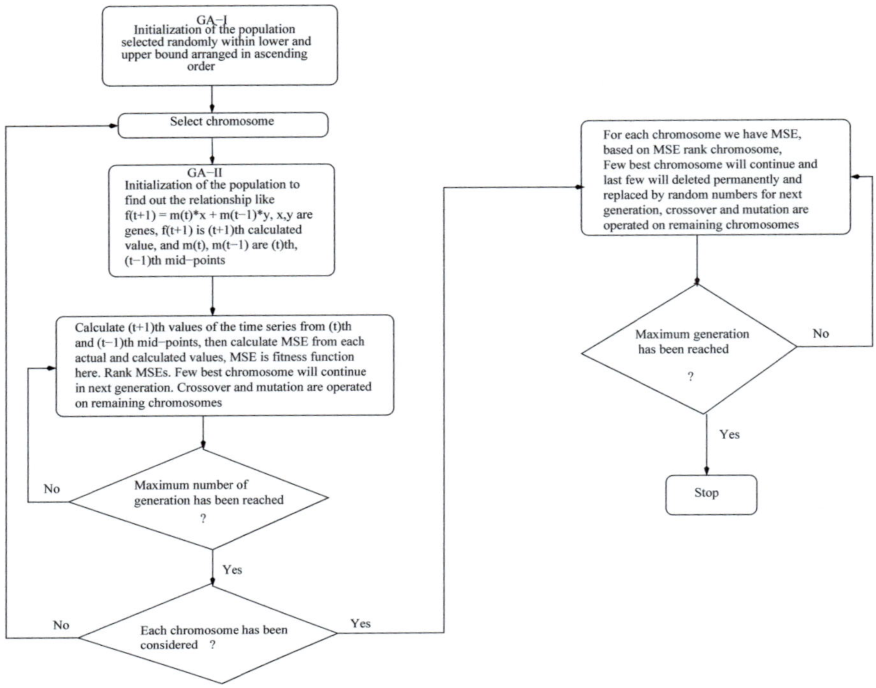

Fig. 1 Flowchart of the proposed methodology

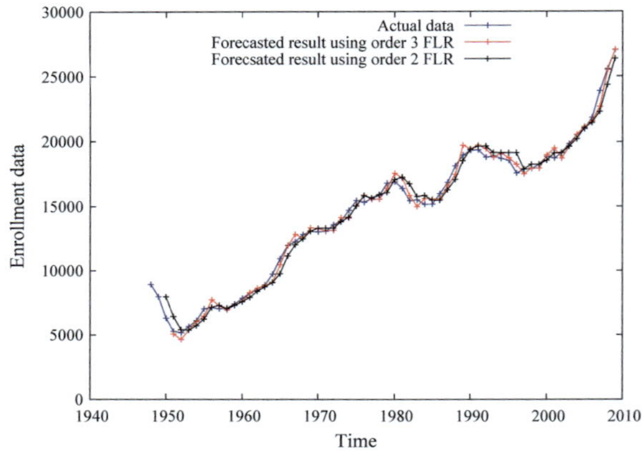

Fig. 2 Fuzzy logical relationship of order 3 and 2 on actual enrollment data

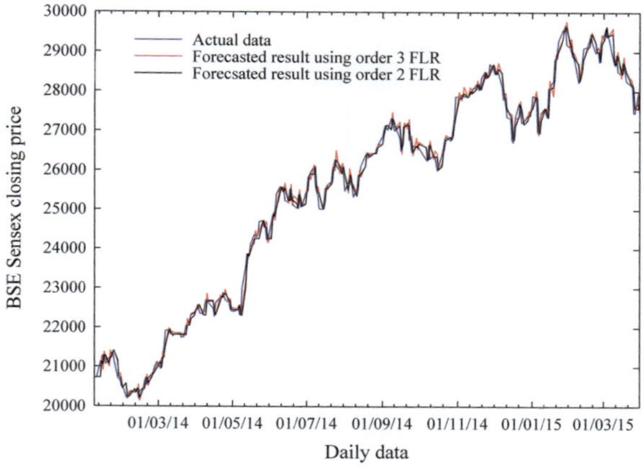

Fig. 3 Fuzzy logical relationship of order 3 and 2 on actual BSE sensex data

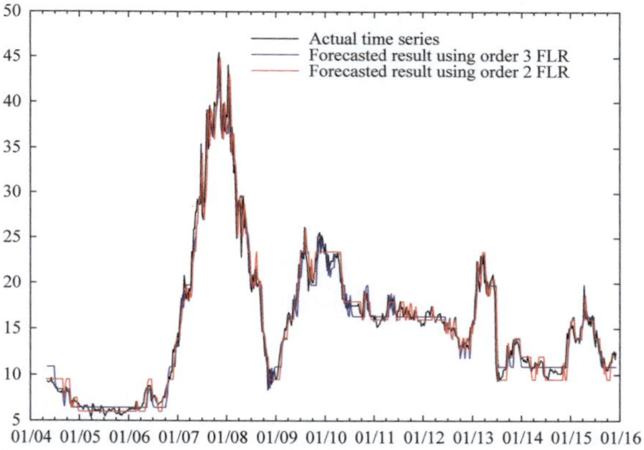

Fig. 4 Fuzzy logical relationship of order 3 and 2 on actual Shenzhen stock exchange data

works. It is observed that our approach outperforms the other conventional models. Here, we have used third-order FLR for the enrollment data set.

Table 2 shows the MSE values corresponding to the three data sets with third-order FLR and second-order FLR. From the calculated MSEs in Table 2, we can say that current situation highly depends on three previous occurrences rather than two previous occurrences in case of enrollment data set, whereas in case of BSE sensex data, current situation highly depends on two previous occurrences rather than three previous occurrences. For Shenzhen stock exchange data, we cannot say anything like this as MSEs are approximately equal in both consideration.

Table 2 specifies the number of genes, population, and generation for GA-I and GA-II, respectively.

Table 1 Performance of conventional models as well as proposed approach on enrollment data set by showing MSE values

Year	Enrollments	Song and Chissom [13]	Song and Chissom [14]	Chen's Method [3]	Hwang et al. [6]	Proposed approach
1948	8,916					
1949	.					
1950	.					
1951	5269					5049.78
.	.					.
.	.					.
.	.					.
1965	10,938					10444.13
1966	11,975					11934.58
1967	12,251					12811.08
1968	12,816					.
1969	13,035					.
1970	13,017					.
1971	13,055					.
1972	13,563	14,000		14,000		13119.32
1973	13,867	14,000		14,000		14118.34
1974	14,696	14,000		14,000		14077.09
1975	15,460	15,500	14,700	15,500		15070.61
1976	15,311	16,000	14,800	16,000	16,260	15877.95
.	.	.	.	16,000	15,511	15526.84
.	.	.	.	.	.	.
.	.	.	.	.	.	.
1990	19,328	19,000	17,800	19,000	19,770	19440.35
1991	19,337	19,000	19,300	19,000	19,928	19732.26
1992	18,876		19,600	19,000	19,537	19451.31
.	.					.
.	.					.
.	.					.
2007	23,878					22651.41
2008	25,580					25493.17
2009	27,052					27087.26
MSE		423,027	775,687	407,507	321,418	185,595.03

Table 2 Number of GA parameters used for each time series with specified order of FLR

GA used for	GA-I			GA-II		
	Genes	Population	Generation	Genes	Population	Generation
Enrollment data FLR order 3	3	2000	10	200	50	5
Enrollment data FLR order 2	2	2000	10	200	50	5
BSE sensex data FLR order 3	3	2000	10	200	50	5
BSE sensex data FLR order 2	2	2000	10	200	50	5
Shenzhen stock, FLR order 3	3	2000	10	20	50	5
Shenzhen stock, FLR order 2	2	2000	10	20	50	5

5 Conclusion

A new model has been proposed for fuzzy time series using double GA to decide the unequal intervals resulted from partitioning the UOD and the fuzzy logical relationship. Here, second-order and third-order fuzzy logic have been considered.

Contributions of this work have been given below.

- This work proposes a new approach to find the relationship among mid-points of consecutive data points which has been used to replace the third step of conventional fuzzy time series model.
- Proposed approach is on establishing the FLRs without matrix operation. Also, it does not suffer from fixed UOD.
- To achieve the above-mentioned purposes, real coded GA has been used here.
- Here, the higher order of FLRs has been also considered.
- Length of interval always effects the forecasting accuracy [7]. To adjust the lengths of intervals by getting unequal length intervals, real coded GA has been developed.
- So, double GA is used here to get the forecasting model.

From result comparisons, it can be concluded that the model performs well.

In future, this model can be improved by searching FLR in different way to get more robust forecasting result.

6 Appendix

If A_i is the ith actual value, F_i is the ith forecasted value, and m denotes the number of historical data, then mean square error (MSE) is given below.

6.1 Mean Square Error (MSE)

If the actual and forecasted values are given, then MSE gives the average of the squares of the errors. Errors are the difference between forecasted and actual value.

$$MSE = \frac{1}{m}\sum_{i=1}^{m}(\mathrm{F}_i - \mathrm{A}_i)^2 \quad (7)$$

References

1. BSE Data Set. https://in.nance.yahoo.com/q/hp?=^BSESN
2. Chen, S.M., Chung, N.Y.: Forecasting enrollments using high-order fuzzy time series and genetic algorithms. Int. J. Intel. Syst. **21**, 485–501 (2006). https://doi.org/10.1002/int.20145
3. Chen, S.M.: Forecasting enrollments based on fuzzy time series. Fuzzy Sets Syst. **81**, 311–319 (1996). https://doi.org/10.1016/0165-0114(95)00220-0
4. Gangwar, S.S., Kumar, S.: Probabilistic and intuitionistic fuzzy sets based method for fuzzy time series forecasting. Cybernet. Syst. Int. J. **45**, 349–361 (2014). https://doi.org/10.1080/01969722.2014.904135
5. Halton, J.: Algorithm 247: radical-inverse quasi-random point sequence. Assoc. Comput. Mach. (ACM) **7**, 701–702 (1964). https://doi.org/10.1145/355588.365104
6. Hwang, J.R., Chen, S.M., Lee, C.H.: Handling forecasting problems using fuzzy time series. Fuzzy Set Syst. **100**, 217–228 (1998). https://doi.org/10.1016/S0165-0114(97)00121-8
7. Huarng, K.: Effective lengths of intervals to improve forecasting in fuzzy time series. Fuzzy Sets Syst. **123**, 387–394 (2001). https://doi.org/10.1016/S0165-0114(00)00057-9
8. Jasim, H.T., Salim, A.G.J., Ibraheem, K.I.: A novel algorithm to forecast enrollment based on fuzzy time series. Appl. Appl. Math. (AAM) **7**, 385–397 (2012)
9. Joshi, B.P., Kumar, S.: Intuitionistic Fuzzy sets based method for fuzzy time series forecasting. Cybernet. Syst. Int. J. **43**, 34–47 (2012). https://doi.org/10.1080/01969722.2012.637014
10. Lee, M.H., Efendi, R., Ismail, Z.: Modified weighted for enrollment forecasting based on Fuzzy time series. Matematika **25**, 67–78 (2009). https://doi.org/10.11113/matematika.v25.n.260
11. Lin, Y., Yang, Y.: Stock markets forecasting based on fuzzy time series model. In: IEEE International Conference Intelligent Computing and Intelligent Systems, pp. 782–786 (2009). https://doi.org/10.1109/ICICISYS.2009.5358026
12. Pal, S.S., Pal, T., Kar, S.: An improvement in forecasting interval based Fuzzy time series. In: Conference Proceedings FUZZ IEEE 2014, Beijing, China, pp. 1390–1394 (2014). https://doi.org/10.1109/FUZZ-IEEE.2014.6891860
13. Song, Q., Chissom, B.S.: Forecasting enrollments with fuzzy time series, Part I. Fuzzy Sets Syst. **54**, 1–9 (1993). https://doi.org/10.1016/0165-0114(93)90355-L
14. Song, Q., Chissom, B.S.: Forecasting enrollments with fuzzy time series, Part II. Fuzzy Sets Syst. **62**, 1–8 (1994). https://doi.org/10.1016/0165-0114(94)90067-1
15. Sakhuja, S., Jain, V., Kumar, S., Chandra, C., Ghildayal, S.K.: Genetic algorithm based fuzzy time series tourism demand forecast model. Indust. Manag. Data Syst. **116**, 483–507 (2016). https://doi.org/10.1108/IMDS-05-2015-0165
16. Singh, P.: Rainfall and financial forecasting using fuzzy time series and neural networks based model. Int. J. Mach. Learn. Cybernet. (2016). https://doi.org/10.1007/s13042-016-0548-5
17. Yu, H.K.: Weighted fuzzy time series models for Taiex forecasting. Phys. A **349**, 609–624 (2005). https://doi.org/10.1016/j.physa.2004.11.006
18. Zhou, R., Zhan, Y., Cai, R., Tong, G.: A mean-variance hybrid-entropy model for portfoli selection with fuzzy returns. Entropy **17**, 3319–3331 (2015). https://doi.org/10.3390/e17053319

A Multi-item EPQ Model with Variable Demand in an Imperfect Production Process

Anindita Kundu, Partha Guchhait, Barun Das and Manoranjan Maiti

Abstract Nowadays, to survive and promote the market competition, multi-item business strategy is more effective for any production/manufacturing sector. Here, an attempt has been made to develop a multi-item production inventory model, especially for seasonal products for a finite time horizon. Production process is not 100% reliable, and for this reason some imperfect quality items are produced, and these items are immediately remanufactured incurring some cost to get back its originality. Both demands and production rate of the items are time-sensitive. It is also assumed that production cost per unit varies with defective rates as well as production rates. Imposing two constraints, space and investment, profit maximization model is formulated. Incorporating inflation and time value of money into the model, total profit is represented as a definite integral with time period as its upper limit. The profit function becomes an optimal control problem. The defective rates of the production process are determined by variational calculus. As today's competitive business transaction is full of uncertainty, another model is considered with fuzzy constraints and solved. Fuzzy constraints are converted to the crisp one, following fuzzy possibility. For the illustration of the developed models, numerical experiments and also some sensitivity analyses are performed and presented.

Keywords EPQ model · Imperfect production process · Time-sensitive production and demand · Variational calculus · Defective rate of the production process

---

A. Kundu · B. Das (✉)
National Institute of Technology Durgapur, Durgapur 713209, WB, India
e-mail: kundua2012@gmail.com

B. Das
e-mail: bdasskbu@gmail.com

P. Guchhait
Vidyasagar University, Midnapore 721102, WB, India

M. Maiti
Sidho-Kanho-Birsha University, Purulia 723104, WB, India
e-mail: mmaiti2005@yahoo.co.in

P. Chandra et al. (eds.), *Information Technology and Applied Mathematics*,
Advances in Intelligent Systems and Computing 699,
https://doi.org/10.1007/978-981-10-7590-2_16

1 Introduction

In a manufacturing system, it is difficult to produce all perfect quality units as a system can never be 100% perfect in the production run-time. Thus as a result, some defective units are produced and repaired instantly. Moreover, the reliability parameter of the machinery system has an impact on the production of imperfect units. In inventory analysis, planning horizon is normally considered as infinite assuming unchanged specification of products forever. This is unrealistic some cases. For the seasonal products—winter garments, Xmas cakes and cards, etc.—the planning horizon is finite. From the market, it is also observed that unexpected improvement of modern technology leads to rapid change in item specifications with more attractive features. For this reason, these types of items normally appear in the market for a short time. Seasonal products, such as hi-tech products like i-Phone, Tab, and also fruits, food grains, normally exist for a certain period of time, and obviously demands of these items depend on it. For almost all the food grains such as rice, potato, dal, demand increases with time. Inflation is continuously raising in the developing countries, and time value of money goes down very sharply. Thus, at present, one cannot ignore the consequences of inflation and time value of money.

Nowadays, in the volatile market, the business of a single item is not safe for a retailer/manufacturer. Rather business of multi-item is preferred as the loss in the business of a particular item may be recovered by the other one. Problems in multi-item business are the resource scarcity; i.e., available money, storage space may not be sufficient and well defined. Again, one of the restrictions of the inventory model that has received much attention in the literature is the impreciseness of the input data. It may occur due to incomplete or unobtainable information. In reality, fuzzy presentation of imprecise input data gets considerable attention from the researchers of inventory control problems.

Considering all these facts into consideration, here a multi-item imperfect EPQ model is developed over a finite time horizon. Defective produced units are remanufactured instantly incurring some cost and treated as a perfect one. The production rate and demands of the items vary with time. The production cost per unit quantity depends on cost for raw material, labor charge, rate of production, and defective rate of the production process. Again, keeping in mind the economic conditions of the developing countries, the model is developed with inflation and time value of money. The model is expressed as an optimal control problem having constraints on space and investment to determine the optimum reliability indicators (r_i) for the maximization of the manufacturer's total profit over a finite planning horizon. This investigation also develops an EPQ model considering the imprecise parameters as fuzzy in nature, which helps the production managers to adopt a better planning strategy to increase his/her profit. The model is illustrated numerically, and the importance of some model's parameters is also outlined.

The novelty of the present paper is given below:

- The main contribution of the paper is that a multi-item EPQ model with two constraints(inequality) is considered and solved using variational principle which is possibly not available in the literature till now.
- In this investigation, an imperfect EPQ model is developed with time-sensitive production and demand.
- Production cost per unit quantity depends on labor charge, cost of raw material, cost of wear and tear, rate of production and defective rates (r_i) of the production process.
- A restricted multi-item imperfect EPQ model (defective rate-dependent) with two constraints has been formulated and solved.
- Another importance of the present investigation is to reveal the effects of defective rates on the production cost per unit quantity and also on the optimal profit.
- In the inventory control problems, very few inventory practitioners have investigated the relation between the defective rates and remanufacturing rate in an optimal control problem.

2 Literature Review

Porteous [21] and Rosenblatt and Lee [23] were the first who discussed process reliability and quality improvement in production inventory models. In their models, they assumed that the imperfect quality units are produced in the out-of-control state of the machinery system. Till now, several EPQ models have been developed by the inventory practitioners with the assumption of process reliability. In the earlier models, it is considered that the very good machineries produce 100% perfect products. Nowadays, manufacturers are highly satisfied if process reliability indicator ('r_i') remains nearer to the minimum value (zero) and they never allow it to become higher than a certain level. In the literature, there are some published articles with this phenomenon, such as Sana [24], Kumar and Goswami [16], and Zhou et al. [29]. Recently, Jaggi et al. [12] presented an excellent article with imperfect production process. For the present investigation, 'r' is assumed as a defective rate of the production process.

Business of multi-items is quite popular [3, 5]; because in the present market conditions, if the business of one particular item brings loss for the retailer, then the other item may save the situation. On the other hand, most of the earlier EPQ models ([21, 26]; etc) are published assuming unchanged specification of the products forever, though the improvement of modern technology motivates the manager of a production sector to change the specifications [11] of the item. Several investigations have already been made by the researchers of inventory control problems along with time-dependent demand [18, 24]. Nowadays, inflation and time value of money is an important tool in any business strategies(mostly in the developing countries). Incorporating this realistic assumption, an EOQ model was developed first by

Buzacott [1]. Recently, inventory practitioners like Sana [24], Guchhait et al. [9], Tiwari et al. [25] gave their effort to develop EPQ/EMQ models with inflation and time value of money. Again, earlier EPQ models are developed with fixed production cost per unit quantity. Khouja [15] first presented an article and proved that the unit production cost may vary with raw material cost, production rate, labor charge, wear and tear cost, and defective rate of the production process.

In 1965, Zadeh first introduced fuzzy set and after that it has been developed in the different branches of science and other areas along with inventory control problems [6–10, 13, 14, 17, 27]. The main advantage of fuzzy set theory lies in its inherent ability to deal with imprecise information based on human judgment. Again, to the best of our knowledge, very few might have used variational principle [7, 24] for the analysis of multi-items imperfect EPQ model with several crisp/uncertain constraints. The main difficulty in treating the optimal control problems having constraints is to connect appropriately the constraints with the functional. As the multi-item inventory problems are always connected with constraints, there are not much EPQ models for multi-items formulate as optimal control problems. This non-availability of multi-item constrained EPQ model motivated us to take the present investigation principle. The Euler–Lagrange equations for the augmented cost functionals are derived and solved by the generalized reduced gradient method using LINGO software.

3 Assumptions and Notations

Notations: (for ith item)

(i) $q_i(t)$ is the inventory level at any time $t(\geq 0)$.
(ii) $\dot{q}_i(t)$ is the derivative of $q_i(t)$ with respect to t.
(iii) $D_i(t)$ is the demand of the i-th item.
(iv) r_i is the defective rate of the production process.
(v) r_{min_i} and r_{max_i} are minimum and maximum value of r_i.
(vi) ξ is the variation constant of tool/die costs.
(vii) c_{d_i} is the rework cost per defective item.
(viii) c_{h_i} is the holding cost per unit.
(ix) c_{3_i} is the setup cost.
(x) s_{p_i} is the unit selling price.
(xi) s_{0_i} is the required space per unit.

Assumptions: (for ith item)

(i) Here, it is assumed that N items are produced.
(ii) Life cycle of the items, i.e., the planning horizon for the model, T is finite.
(iii) For seasonal/fashionable products, it is well known that the business period is finite, yet it depends on time. Here, demand rate is assumed to be time-dependent, $D_i(t)$ for ith item.

(iv) Production rate also increases with time and is of the form $P_i(t)$ for the ith item.

(v) $\mho(r_i)$ is the development cost of production and represented as (cf. [20] and [24]):

$$\mho(r_i) = \mho_{A_i} + \mho_{B_i} e^{\omega_i (r_{max_i} - r_i)/(r_i - r_{min_i})},$$

where r_i is the defective rates of the production process, $\mho_{A_i}$ is the fixed cost like labor, energy. $\mho_{B_i}$ represents the technology cost, resource and design complexity for production when $r_i = r_{max_i}$. ω_i represents the difficulties in increasing reliability, which depends on the design complexity, technology and resource limitations, etc.

(vi) Unit production cost, c_{p_i}, is a function of defective rate r_i and production rate $P_i(t)$ and is of the form

$$c_{p_i}(r_i, t) = c_{r_i} + \frac{\mho(r_i)}{P_i(t)} + \xi_i \, P_i(t).$$

where c_{r_i} is the fixed material cost. Second term is the development cost which is equally distributed over the production $P_i(t)$ at any time t, and the third term $\xi_i P_i(t)$ is tool/die cost which is proportional to the production rate.

(vii) In the developing countries, inflation is predominant and interest rate depends on the inflation value. Thus $\delta = R - i$, where R and i are the interest and inflation per unit currency, respectively.

4 Model Formulation

Here, the production system produces fresh units and at the same time some defective units also. Produced defective units are reworked incurring some cost and treated as perfect ones. In this production process, r_i plays the role as a defective rate of the production process. So, smaller value of r_i means production process produces better quality product. Thus, the EPQ system is represented as:

$$\frac{dq_i(t)}{dt} = P_i - D_i,$$
$$\text{i.e., } \mathrm{P_i} = \dot{q_i} + D_i, \quad i = 1, ..., N \tag{1}$$
$$\text{with } q_i(0) = 0 \text{ and } q_i(T) = 0$$

where D_i and P_i stand for $D_i(t)$ and $P_i(t)$, respectively.

Now, the total profit during the time period $[0, T]$ (using Eq. 1) along with inflation and time value of money can be obtained from the following equation:

$$Z = \int_0^T e^{-\delta t} \sum_{i=1}^{N} [s_{p_i} D_i(t) - c_{p_i}(r,t) P_i(t) - c_{d_i} r_i P_i(t) - c_{h_i} q_i(t) - c_{3_i}] \tag{2}$$

Here, we have to find out the values of $q_i(t)$, which maximize the total profit within the interval [0, T], and obviously, $q_i(t)$ must satisfy the following two constraints

$$\sum_{i=1}^{N} (s_{0_i} q_i) \leq S_1 \tag{3}$$

$$\text{and} \quad \sum_{i=1}^{N} (c_{r_i} q_i) \leq CB_1 \tag{4}$$

Which can be written as

$$\sum_{i=1}^{N} (s_{0_i} q_i) = S \ \text{ where } 0 \leq S \leq S_1 \quad \text{and}$$

$$\sum_{i=1}^{N} (c_{r_i} q_i) = CB \ \text{ where } 0 \leq CB \leq CB_1$$

Here, we formulate the augmented cost functional by introducing the Lagrange's multiplier λ and μ as follows:

$$Z = \int_0^T e^{-\delta t} \left[\sum_{i=1}^{N} [(s_{p_i} - c_{r_i} - c_{d_i} r_i)(\dot{q}_i + D_i) - \mho(r_i) - \xi_i(\dot{q}_i + D_i)^2 - c_{h_i} q_i - c_{3_i}] + \lambda[\sum_{i=1}^{N} (s_{0_i} q_i) - S] + \mu[\sum_{i=1}^{N} (c_{r_i} q_i) - CB] \right] dt = \int_0^T \Phi(q_i, \dot{q}_i, t) dt, \tag{5}$$

$$\text{where} \quad \Phi(q_i, \dot{q}_i, t) = e^{-\delta t} \left[\sum_{i=1}^{N} [(s_{p_i} - c_{r_i} - c_{d_i} r_i)(\dot{q}_i + D_i) - \mho(r_i) - xi_i(\dot{q}_i + D_i)^2 - c_{h_i} q_i - c_{3_i}] + \lambda[\sum_{i=1}^{N} (s_{0_i} q_i) - S] + \mu[\sum_{i=1}^{N} (c_{r_i} q_i) - CB] \right] \tag{6}$$

For the optimal value of Z, the necessary condition is that Z must satisfy the Euler–Lagrange equation. So, we have to find out the values of $q(t)$ and also $P(t)$ for which Z can be maximized.

Hence, Euler–Lagrange's equation is

$$\frac{\partial \Phi}{\partial q_i} - \frac{d}{dt}\left(\frac{\partial \Phi}{\partial \dot{q}_i}\right) = 0. \tag{7}$$

Using (5) and (6), we have

$$\ddot{q}_i - \delta \dot{q}_i = \Psi(t) \tag{8}$$

where

$$\Psi(t) = \delta D_i - \dot{D}_i + c_{h_i}/(2\xi_i) - (\delta/2\xi)$$
$$(s_{p_i} - c_{r_i} - c_{d_i} r_i) - (\lambda s_{0_i} + \mu c_{r_i}) e^{\delta t}/2\xi_i$$

Now, we have to solve Eq. (8) to get the expression for $q_i(t)$ and that of $P_i(t)$. For solution, the complementary function (CF) of Eq. (8) is given by

$$q_i(t) = X_{1i} + X_{2i} e^{\delta t}, \qquad i = 1, ..., N \tag{9}$$

where X_{1i} and X_{2i} are the arbitrary constants. Again, particular integral (PI) of Eq. (8) is given by

$$\frac{1}{D^2 - \delta D}[\Psi(t)] \quad [\text{Here } D(\equiv \frac{d}{dt})] \tag{10}$$

It can be investigated from today's world market that the demand of many products especially hi-tech products increases significantly during the growth period. For this reason, it is considered that the demand functions of the items are exponentially time-dependent and are of the form

$$D_i(t) = a_i e^{b_i t},$$

where a_i and b_i are constants.
Therefor, the PI of Eq. (8) is given by

$$\text{PI} = X_{3i} e^{b_i t} + X_{4i} t e^{\delta t} + X_{5i}(t + \frac{1}{\delta}) \tag{11}$$

where $X_{3i} = -a_i/b_i$, $X_{4i} = -(\lambda s_{0_i} + \mu c_{r_i})/(2\xi_i \delta)$, and $X_{5i} = -\{c_{h_i} - \delta(s_{p_i} - c_{r_i} - c_{d_i} r_i)\}/(2\xi_i \delta)$. Hence, the complete solution of Eq. (8) is CF+PI, i.e.,

$$q_i(t) = X_{1i} + X_{2i}e^{\delta t} + X_{3i}e^{b_i t} + X_{4i}te^{\delta t} + X_{5i}(t + \frac{1}{\delta}) \tag{12}$$

As $q(0) = 0 = q(T)$, we can get the value of the constants X_{1i} and X_{2i}. Also, as

$$P_i(t) = \dot{q}_i(t) + D_i(t)$$

which implies,

$$P_i(t) = \delta X_{2i}e^{\delta t} + b_i X_{3i}e^{b_i t} + X_{4i}(\delta te^{\delta t} + e^{\delta t}) + X_{5i} + a_i e^{b_i t} \tag{13}$$

From Eqs. (12) and (13), putting the value of $q(t)$ and $P(t)$, respectively, in Eq. (5), it can evaluate the expression for $\Phi(q_i, \dot{q}_i, t)$ and hence also the profit function Z can be obtained. Here, the profit function Z is given by

$$\begin{aligned}
Z = \sum_{i=1}^{N} \Bigg[& (s_{p_i} - c_{r_i} - c_{d_i} r_i)\{\delta X_{2i}T + X_{4i}(T + \delta T^2/2) + \frac{(X_{3i}b_i + a_i)(e^{(b_i-\delta)T})}{(b_i - \delta)} \\
& + X_{5i}(1 - e^{-\delta T})/\delta\} - \mho(1 - e^{-\delta T})/\delta - \xi[\delta^2 X_{2i}^2(e^{\delta T} - 1)/\delta + 2\delta X_{2i}X_{4i} \\
& \{(e^{\delta T} - 1)/\delta + (\delta Te^{\delta T} - e^{\delta T} + 1)/\delta\} + X_{4i}^2\{(e^{\delta T} - 1)/\delta + 2(\delta Te^{\delta T} - \\
& e^{\delta T} + 1)/\delta + (\delta^2 T^2 e^{\delta T} - 2\delta Te^{\delta T} + e^{\delta T} - 1)/\delta\} + (X_{3i}b_i + a_i)^2 \frac{1}{(2b_i - \delta)} \\
& (e^{(2b_i-\delta)T} - 1) + X_{5i}^2(1 - e^{-\delta T})/\delta + 2(X_{3i}b_i + a_i)\{(\delta X_{2i} + X_{4i})\frac{e^{b_i T} - 1}{b_i} \\
& + X_{4i}(\delta Te^{b_i T} - e^{b_i T} + 1)/\delta\} + 2X_{5i}(\delta X_{2i}T + X_{4i}T + X_{4i}\delta T^2/2) \\
& + 2X_{5i}(X_{3i}b_i + a_i)(e^{(b_i-\delta)T} - 1)/(b_i - \delta)] - c_{h_i}\{(X_{1i} + X_{5i}/\delta) \\
& (1 - e^{-\delta T})/\delta + X_{2i}T + X_{3i}(e^{(b_i-\delta)T} - 1)/(b_i - \delta) + \frac{X_{4i}T^2}{2} - X_{5i}(\delta Te^{-\delta T} \\
& + e^{-\delta T} - 1)/\delta^2\} - c_{3_i}(1 - e^{-\delta T})/\delta + [(\lambda s_{0_i} + \mu c_{r_i})\{(X_{1i} + X_{5i}/\delta)T \\
& + (X_{2i}/\delta)(e^{\delta T} - 1) + (X_{3i}/\delta)(e^{b_i T} - 1) + X_{5i}T^2/2 + (X_{4i}/\delta)(Te^{\delta T}/\delta \\
& - e^{\delta T}/\delta^2 + 1/\delta)\} - \lambda ST - \mu CBT]\Bigg]
\end{aligned} \tag{14}$$

Here, for the optimum values of Z, λ, and μ are selected by trial and error method. As the profit function contains two variables r_1 and r_2, it is optimized with respect to r_1 and r_2 using a nonlinear gradient-based technique—generalized reduced gradient (GRG) technique [22] by LINGO software.

4.1 Fuzzy Model (FM)

Keeping in mind the impreciseness of the input data, available space, $\widetilde{S} = (S_1, S_2, S_3)$, and the budget, $\widetilde{CB} = (CB_1, CB_2, CB_3)$, are assumed as triangular fuzzy number; i.e., both the constraints are fuzzy in nature. So, the above problem can be represented as:

$$\left.\begin{array}{l} \text{Determine } \; r_i, i = 1, 2, ...N \text{ to} \\ \text{Maximize } \; Z \\ \text{subject to } \; \sum_{i=1}^{N} (s_{0_i} q_i) \leq \widetilde{S}, \\ \text{and } \; \sum_{i=1}^{N} (c_{r_i} q_i) \leq \widetilde{CB} \\ \text{where } Z \text{ is given by Eq. (2).} \end{array}\right\} \tag{15}$$

Here, we use fuzzy possibility to convert the fuzzy constraints of the above problem into crisp one which is represented as follows:

$$\left.\begin{array}{l} \text{Determine } \; q_i, i = 1, 2, ...N \text{ to} \\ \text{Maximize } \; Z \\ \text{subject to } \; \text{pos}(\sum_{i=1}^{N} (s_{0_i} q_i) \leq \widetilde{S}) \geq \beta_1 \\ \text{and } \qquad \text{pos}(\sum_{i=1}^{N} (c_{r_i} q_i) \leq \widetilde{CB}) \geq \beta_2 \\ \text{where } Z \text{ is given by Eq. (2).} \end{array}\right\} \tag{16}$$

where ‘pos’ represents possibility measure of fuzzy number and β_1 and β_2 are pre-defined possibility levels. Using the Lemma-3 [5], the problem reduces to

$$\left.\begin{array}{l} \text{Determine } \; q_i, i = 1, 2, ...N \text{ to} \\ \text{Maximize } \; Z \\ \text{subject to } \; \sum_{i=1}^{N} (s_{0_i} q_i) \leq (1 - \beta_1) S_3 + \beta_1 S_2, \\ \text{and } \qquad \sum_{i=1}^{N} (c_{r_i} q_i) \leq (1 - \beta_2) CB_3 + \beta_2 CB_2 \\ \text{where } Z \text{ is given by Eq. (2).} \end{array}\right\} \tag{17}$$

In this case also, the augmented cost functional (Z_F, say) can be formulated by introducing Lagrange’s multiplier λ and μ as follows:

$$Z_F = \int_0^T e^{-\delta t} \left[\sum_{i=1}^{N} [(s_{p_i} - c_{r_i} - c_{d_i} r_i)(\dot{q_i} + D_i) \right.$$

$$-\mho(r_i) - \xi_i(\dot{q}_i + D_i)^2 - c_{h_i}q_i - c_{3_i}] +$$
$$\lambda[\sum_{i=1}^{N}(s_{0_i}q_i) - \{(1-\beta_1)S_3 + \beta_1 S_2\}] +$$
$$\left.\mu[\sum_{i=1}^{N}(c_{r_i}q_i) - \{(1-\beta_2)CB_3 + \beta_2 CB_2\}]\right] dt$$
$$= \int_0^T \Phi_F(q_i, \dot{q}_i, t)dt, \tag{18}$$

where

$$\Phi_F(q_i, \dot{q}_i, t) = e^{-\delta t}\left[\sum_{i=1}^{N}[(s_{p_i} - c_{r_i} - c_{d_i}r_i)\right.$$
$$(\dot{q}_i + D_i) - \mho(r_i) - \xi_i(\dot{q}_i + D_i)^2 - c_{h_i}q_i - c_{3_i}]$$
$$+\lambda[\sum_{i=1}^{N}(s_{0_i}q_i) - \{(1-\beta_1)S_3 + \beta_1 S_2\}]$$
$$\left.+\mu[\sum_{i=1}^{N}(c_{r_i}q_i) - \{(1-\beta_2)CB_3 + \beta_2 CB_2\}]\right] \tag{19}$$

Proceeding as before, we can get the profit function Z_F as follows:

$$Z = \sum_{i=1}^{N}\left[\left[(s_{p_i} - c_{r_i} - c_{d_i}r_i)\{\delta X_{2i}T + X_{4i}(T + \delta T^2/2) + \frac{(X_{3i}b_i + a_i)(e^{(b_i-\delta)T})}{(b_i - \delta)}\right.\right.$$
$$+X_{5i}(1 - e^{-\delta T})/\delta\} - \mho(1 - e^{-\delta T})/\delta - \xi[\delta^2 X_{2i}^2(e^{\delta T} - 1)/\delta + 2\delta X_{2i}X_{4i}$$
$$\{(e^{\delta T} - 1)/\delta + (\delta T e^{\delta T} - e^{\delta T} + 1)/\delta\} + X_{4i}^2\{(e^{\delta T} - 1)/\delta + 2(\delta T e^{\delta T} - e^{\delta T}$$
$$+1)/\delta + (\delta^2 T^2 e^{\delta T} - 2\delta T e^{\delta T} + e^{\delta T} - 1)/\delta\} + (X_{3i}b_i + a_i)^2(1/(2b_i - \delta))$$
$$(e^{(2b_i-\delta)T} - 1) + X_{5i}^2(1 - e^{-\delta T})/\delta + 2(X_{3i}b_i + a_i)\{(\delta X_{2i} + X_{4i})\frac{(e^{b_i T} - 1)}{b_i}$$
$$+X_{4i}(\delta T e^{b_i T} - e^{b_i T} + 1)/\delta\} + 2X_{5i}(\delta X_{2i}T + X_{4i}T + X_{4i}\delta T^2/2)$$
$$+2X_{5i}(X_{3i}b_i + a_i)(e^{(b_i-\delta)T} - 1)/(b_i - \delta)] - c_{h_i}\{(X_{1i} + X_{5i}/\delta)$$
$$(1 - e^{-\delta T})/\delta + X_{2i}T + X_{3i}(e^{(b_i-\delta)T} - 1)/(b_i - \delta) + X_{4i}T^2/2 - X_{5i}(\delta T$$
$$e^{-\delta T} + e^{-\delta T} - 1)/\delta^2\} - c_{3_i}(1 - e^{-\delta T})/\delta + [(\lambda s_{0_i} + \mu c_{r_i})$$
$$\{(X_{1i} + X_{5i}/\delta)T + (X_{2i}/\delta)(e^{\delta T} - 1) + (X_{3i}/\delta)(e^{b_i T} - 1) + X_{5i}T^2/2$$
$$+(X_{4i}/\delta)(Te^{\delta T}/\delta - e^{\delta T}/\delta^2 + 1/\delta)\} - \lambda\{(1-\beta_1)S_3 + \beta_1 S_2\}T$$
$$-\mu\{(1-\beta_2)CB_3 + \beta_2 CB_2\}T]] \tag{20}$$

5 Numerical Experiment

For the illustration of the models numerically, it is assumed that the production sector produces two items, i.e., $N = 2$.

The parametric values used to illustrate the crisp model are presented in Table 1.

Other parametric values such as $CB = 7850$, $S = 500$, $T = 2$, and $\delta = 0.01$ are used to illustrate the crisp model. Now, with these set of parametric values, obtained results are presented in Table 2.

From the obtained results, it is reflected that, with the initial value of λ and μ, the budget constraint of the model is not satisfied. After a certain decreasing and increasing value of λ and μ, respectively, both the constraints are satisfied and the obtained profit is more than the others. Again, it is found that, though the constraints are satisfied, the profit of the model is less than the profit for the values of $\lambda = 1.50$ and $\mu = 0.597$. Initially, with the decreasing and increasing values of λ and μ, respectively, budget constraints of the problem are not satisfied. It may effect on the profit function of the model. It is observed from Table 2 that the profit reflects maximum value for $\lambda = 1.50$ and $\mu = 0.597$. For this reason, further investigation is made with these values of λ, μ.

Table 1 Parametric values for crisp model

$item(i)$	a_i	b_i	ξ_i	c_{r_i}	c_{d_i}	c_{3_i}	c_{hi}	w_i	s_{p_i}	$\mho_{A_i}$	$\mho_{B_i}$	s_{0_i}	r_{max_i}	r_{min_i}
1	46	2	0.010	5.0	4.0	90	2.0	0.20	20	5	15	0.5	0.90	0.2
2	40	2	0.011	6.5	4.2	110	2.1	0.25	25	6	10	0.4	0.99	0.2

Table 2 Obtained results for crisp model

λ	μ	r_1	r_2	CB	S	Profit
1.70	0.570	0.2724	0.2835	7882.24	496.42	19977.29
1.65	0.570	0.2733	0.2844	7874.11	496.14	19984.50
1.60	0.580	0.2761	0.2851	7861.39	495.68	19991.70
1.57	0.580	0.2736	0.2823	7859.13	495.30	19996.07
1.55	0.590	0.2711	0.2861	7856.94	495.19	19997.66
1.53	0.592	0.2721	0.2834	7855.51	494.87	19998.80
1.51	0.595	0.2714	0.2843	7853.43	494.52	20001.70
1.50	0.595	0.2717	0.2829	7852.45	494.63	20004.84
1.50	**0.597**	**0.2710**	**0.2844**	**7849.69**	**494.88**	**20006.02**
1.48	0.600	0.2722	0.2847	7848.85	494.54	20003.67
1.46	0.600	0.2725	0.2866	7848.32	495.12	20001.53
1.45	0.610	0.2715	0.2859	7848.01	496.22	19998.18
1.43	0.610	0.2738	0.2845	7847.91	496.87	19997.62

Table 3 Results of the model for different values of r_1 and r_2

r_1	r_2	Profit	r_1	r_2	Profit	r_1	r_2	Profit	r_1	r_2	Profit
0.24		20001.15	0.24		20003.51	0.24		20002.62	0.24		20000.46
0.30		20005.18	0.30		20005.93	0.30		20002.93	0.30		20000.21
0.40		20002.87	0.40		20003.32	0.40		20002.07	0.40		19999.17
0.50	0.24	20000.45	0.50	0.30	20000.76	0.50	0.40	19999.23	0.50	0.50	19997.49
0.60		19997.72	0.60		19997.21	0.60		19996.02	0.60		19994.19
0.70		19995.87	0.70		19995.08	0.70		19993.34	0.70		19991.03
0.80		19994.08	0.80		19993.01	0.80		19991.05	0.80		19988.09
0.90		19992.53	0.90		19991.51	0.90		19989.02	0.90		19985.22
0.24		19998.77	0.24		19995.76	0.24		19992.01	0.24		19987.12
0.30		19997.88	0.30		19994.95	0.30		19991.02	0.30		19985.34
0.40		19996.27	0.40		19993.09	0.40		19989.15	0.40		19981.28
0.50	0.60	19993.91	0.50	0.70	19989.74	0.50	0.80	19985.24	0.50	0.9	19975.46
0.60		19990.48	0.60		19985.64	0.60		19979.91	0.60		19968.33
0.70		19986.62	0.70		19981.28	0.70		19973.98	0.70		19960.78
0.80		19982.27	0.80		19976.33	0.80		19968.48	0.80		19952.98
0.90		19977.79	0.90		19971.12	0.90		19962.42	0.90		19945.05

Results for different values of r_1 and r_2: To analyze the model perfectly, results for different values of r_1 and r_2(i.e., the defective rates of the production process) are obtained and presented in Table 3.

From this table, it is found that with the primary value of r_1 (when r_2 is fixed lower value), profit of the model increases, and after reaching a maximum value, profit decreases with the increasing value of r_1. Again, profit of the model reflects as a decreasing function with the increasing value of r_1 for the fixed value of r_2 (for $r_2 = 0.5$ and higher). To support our study, results are presented graphically. From Fig. 1, it is clearly seen that it is concave in nature. Profit function of the model initially increases with the increasing value of r_1 and r_2; obtaining maximum profit for the particular value of r_1 and r_2, it decreases. Also, it is observed that the profit function has only one pick point within the given range of r_1 and r_2, and it can be concluded that there is a maximum profit for this model within the time interval.

Again from Fig. 2, it is observed that for the lower value of r_2, profit function is concave and concavity of the function becomes less with the higher value of r_2. Also for $r_1 = 0.2$ and $r_2 = 0.3$, the profit function has a pick point and it is clear that there is a maximum profit when r_1 varies from 0.2 to 0.9. But after the fixed value of $r_2 = 0.4$, the profit function has no maximum point for the value of r_1 (0.2–0.9).

Today's competitive volatile market situation is full of uncertainties. Keeping this in mind, another submodel is developed with the consideration of some inventory parameters as fuzzy in nature. In this model, it considered that both the constraints are imprecise in nature. Results of the model are presented in Table 4.

Fig. 1 Profit against the two variables r_1 and r_2

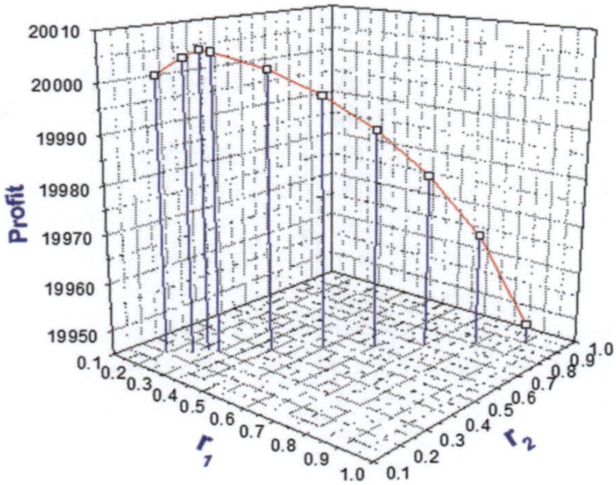

Fig. 2 Graphical representation of the crisp model

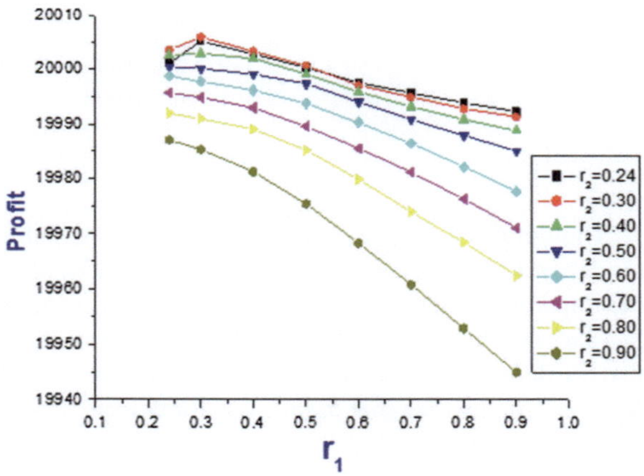

Table 4 Obtained results for fuzzy model (FM)

Model	r_1	r_2	Profit
FM	0.2765	0.2881	20002.21

Results of Fuzzy Model: For the fuzzy model, $\widetilde{CB} = (7840, 7849, 7855)$, $\tilde{S} = (490, 501, 505)$, $T = 2$ and $\delta = 0.01$ are used to illustrate the model. The possibility label of the constraints is taken as $\beta_1 = 0.6$ and $\beta_2 = 0.7$, respectively. Profit is maximum for $\lambda = 1.50$ and $\mu = 0.597$, and it is observed from Table 1. For this reason, with these values of λ, μ and the parametric values mentioned above, the results of the fuzzy model are obtained and presented in Table 4.

Table 5 Results of the FM for different values of r_1 and r_2

r_1	r_2	Profit	r_1	r_2	Profit	r_1	r_2	Profit	r_1	r_2	Profit
0.24		19999.12	0.24		19999.78	0.24		19998.62	0.24		19997.55
0.30		20001.01	0.30		20001.57	0.30		19998.05	0.30		19996.51
0.40		20000.23	0.40		20000.62	0.40		19997.11	0.40		19995.07
0.50	0.24	19999.06	0.50	0.30	19998.49	0.50	0.40	19995.84	0.50	0.50	19993.22
0.60		19997.37	0.60		19996.58	0.60		19993.99	0.60		19991.07
0.70		19995.41	0.70		19994.51	0.70		19992.16	0.70		19988.47
0.80		19993.16	0.80		19992.29	0.80		19989.44	0.80		19985.73
0.90		19990.53	0.90		19989.88	0.90		19987.02	0.90		19982.04
0.24		19995.91	0.24		19993.04	0.24		19989.21	0.24		19985.42
0.30		19994.23	0.30		19991.72	0.30		19988.04	0.30		19983.34
0.40		19992.54	0.40		19989.44	0.40		19985.26	0.40		19978.83
0.50	0.60	19990.12	0.50	0.70	19986.69	0.50	0.80	19981.65	0.50	0.9	19972.89
0.60		19987.16	0.60		19983.08	0.60		19976.99	0.60		19966.07
0.70		19983.08	0.70		19979.02	0.70		19971.02	0.70		19959.09
0.80		19979.3	0.80		19973.21	0.80		19964.75	0.80		19951.65
0.90		19975.07	0.90		19968.07	0.90		19957.08	0.90		19943.54

Though it is not possible to compare the results of imprecise model with the crisp model, yet it can be said that the profit (20006.02$) in crisp model is more than the imprecise mode l (20002.21$). In the fuzzy model, for the profit function, many factors may play their important role. From the tables, i.e., Tables 2 and 4, it observed that the defective rate of the production process, i.e., the values of r_1 and r_2, are more for fuzzy model than that of the crisp model and profit can be effected by the smaller values of r_1 and r_2.

Being influenced on the crisp model and to analyze the imprecise model, the results of the fuzzy model are also obtained for different values of r_1 and r_2 and are presented in Table 5. From this table, it is found that the nature of the profit function is similar to the crisp model. To support our study results of fuzzy model is also presented graphically (cf. Fig. 3). From the figure, it is observed that for preceding value of r_2, profit function has maximum profit and it is also observed that the nature of the graph in the fuzzy model is similar to the crisp ones.

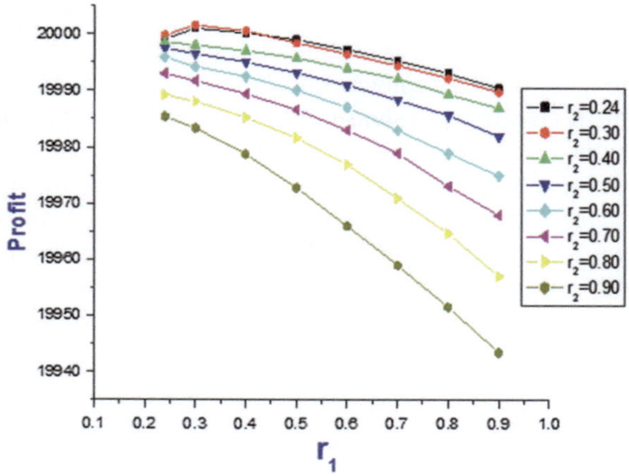

Fig. 3 Graphical representation of the fuzzy model

6 Practical Uses

Here, different models are formulated and solved with respect to a production manager under crisp and fuzzy environments along with defective rates as a decision variable. Owner of any production sector may take the managerial decisions which are discussed below.

(i) Nowadays, business of multi-item is preferred as the loss in the business of a particular item may be recovered by the other one. From the developed multi-item inventory models, management of any production sector may choose the particular model.
(ii) After fixing up the particular model and to achieve maximum profit, manager and/or decision maker himself/herself can take decision about the reduction of defective rates of the production process.

7 Conclusion

In this article, we have studied a multi-item imperfect EPQ inventory model with a fixed time cycle. The model is developed with time-sensitive production rate and demand. Here, in the imperfect production process, defective rate and production rate both are controllable. Again, the developed EPQ model is solved following variational principle. It is a simple technique to optimize a function appearing in a definite integral. From the present investigation, it reveals that the defective rate of the production process should be considered seriously as the production rate is determined by this, and consequently, the production cost per unit and optimal profit are determined by it. This factor should be taken into account by the production manager. It is also observed that the high reliability of production rate does not give the maximum profit always. From the present investigation, considering imprecise inventory costs (fuzzy constraints), another submodel has been developed and solved. In fuzzy model, imprecise constraints are converted to the crisp constraints following fuzzy possibility method and the nature of the obtained results for this model is similar to the crisp model.

From the present investigation, two warehouses EPQ/EOQ model can be modeled, which is more appropriate at present. The model can be developed for deteriorating items with different demand functions. For deteriorating items, preservation technology cost may be imposed to reduce deterioration of the items. Other environments like random, rough, random-rough can be considered for the future extension of the model.

References

1. Buzacott, J.A.: Economic order quantities with inflation. Operat. Res. Quart. **26**, 553–558 (1975)
2. Charnes, A., Cooper, W.W.: Chance constrained programming. Manag. Sci. **6**, 73–79 (1959)
3. Cardenas-Barron, L.E., Sana, S.S.: Multi-item EOQ inventory model in a two-layer supply chain while demand varies with a promotional effort. Appl. Math. Model. **39**(21), 6725–6737 (2015)
4. Dubois, D., Prade, H.: Fuzzy sets and system—theory and application. Academic, New York (1980)
5. Guchhait, P., Maiti, M.K., Maiti, M.: Multi-item inventory model of breakable items with stock-dependent demand under stock and time dependent breakability rate. Comput. Indust. Eng. **59**, 911–920 (2010)
6. Guchhait, P., Maiti, M.K., Maiti, M.: Two storage inventory model of a deteriorating item with variable demand under partial credit period. Appl. Soft Comput. **13**, 428–448 (2013)
7. Guchhait, P., Maiti, M.K., Maiti, M.: Production-inventory models for a damageable item with variable demands and inventory costs in an imperfect production process. Int. J. Prod. Econom. **144**, 180–188 (2013)
8. Guchhait, P., Maiti, M.K., Maiti, M.: Inventory model of a deteriorating item with price and credit linked fuzzy demand: a fuzzy differential equation approach **51**(3), 321–353 (2014)
9. Guchhait, P., Maiti, M.K., Maiti, M.: Inventory policy of a deteriorating item with variable demand under trade credit period. Comput. Indust. Eng. **76**, 75–88 (2014)
10. Guchhait, P., Maiti, M.K., Maiti, M.: An EOQ model of deteriorating item in imprecise environment with dynamic deterioration and credit linked demand. Appl. Math. Model. **39**, 6553–6567 (2015)
11. Gurnani, C.: Economic analysis of inventory systems. Int. J. Prod. Res. **21**, 261–277 (1983)
12. Jaggi, C.K., Cardenas-Barron, L.E., Tiwari, S., Shafi, A.: Two-warehouse inventory model for deteriorating items with imperfect quality under the conditions of permissible delay in payments. Scientia Iranica **24**(1), 390–412 (2017)
13. Jana, D.K., Pramanik, S., Maiti, M.: Mean and CV reduction methods on Gaussian type-2 fuzzy set and its application to a multilevel profit transportation problem in a two-stage supply chain network. Neural Comput. Appl. **28**(9), 2703–2726 (2017)
14. Kazemi, N., Olugu, E.U., Abdul-Rashid, S.H., Ghazilla, R.A.B.R.: Development of a fuzzy economic order quantity model for imperfect quality items using the learning effect on fuzzy parameters. J. Intel. Fuzzy Syst. **28**(5), 2377–2389 (2015)
15. Khouja, M.: The economic production lot size model under volume flexibility. Comput. Oper. Res. **22**, 515–525 (1995)

16. Kumar, R.S., Goswami, A.: A fuzzy random EPQ model for imperfect quality items with possibility and necessity constraints. Appl. Soft Comput. **34**, 838–850 (2015)
17. Kundu, A., Guchhait, P., Panigrahi, G., Maiti, M.: An imperfect EPQ model for deteriorating items with promotional effort dependent demand. J. Intel. Fuzzy Syst. **33**, 649–666 (2017)
18. Lee, C.C., Hsu, S.L.: A two-warehouse production model for deteriorating inventory items with time-dependent demands. Eur. J. Operat. Res. **194**(3), 700–710 (2009)
19. Maiti, M.K., Maiti, M.: Production policy for damageable items with variable cost function in an imperfect production process via genetic algorithm. Math. Comput. Model. **42**, 977–990 (2005)
20. Mettas, A.: Reliability allocation and optimization for complex systems. In: Proceedings of the Annual Reliability and Maintainability Symposium. Institute Electrical and Electronics Engineers, Piscataway, NJ, pp. 216–221 (2000)
21. Porteous, E.L.: Optimal lot sizing, process quality improvement and set-up cost reduction. Operat. Res. **34**, 137–144 (1986)
22. Rao, S.S.: Engineering Optimization, Theory and Practice. New age International (P) Limited, India (2000)
23. Rosenblatt, M.J., Lee, H.L.: Economic production cycles with imperfect production processes. IIE Trans. **17**, 48–54 (1986)
24. Sana, S.S.: A production-inventory model in an imperfect production process. Eur. J. Operat. Res. **200**, 451–464 (2010)
25. Tiwari, S., Cardenas-Barron, L.E., Khanna, A., Jaggi, C.K.: Impact of trade credit and inflation on retailer's ordering policies for non-instantaneous deteriorating items in a two-warehouse environment. Int. J. Prod. Econom. **176**, 154–169 (2016)
26. Wu, C., Zhao, Q.: Two retailer supplier supply chain models with default risk under trade credit policy. Springer Plus **5**, 1728 (2016). https://doi.org/10.1186/s40064-016-3346-3
27. Yano, H.: Multiobjective programming problems involving fuzzy coefficients, random variable coefficients and fuzzy random variable coefficients. Int. J. Uncertain. Fuzziness Knowledg. Based Syst. **23**(04), 483–504 (2015)
28. Yoo, S.H., Kim, D., Park, M.S.: Economic production quantity model with imperfect-quality items, two-way imperfect inspection and sales return. Int. J. Prod. Econom. **121**, 255–265 (2009)
29. Zhou, Y.W., Chen, J., Wu, Y.: EPQ models for items with imperfect quality and one-time-only discount. Appl. Math. Model. **39**, 1000–1018 (2015)
30. Zadeh, L.A.: Fuzzy sets. Informat. Control **8**, 338–356 (1965)

Author Index

P. Chandra et al. (eds.), *Information Technology and Applied Mathematics*,
Advances in Intelligent Systems and Computing 699,
https://doi.org/10.1007/978-981-10-7590-2

MIX
Papier aus verantwortungsvollen Quellen
Paper from responsible sources
FSC® C105338

If you have any concerns about our products,
you can contact us on
ProductSafety@springernature.com

In case Publisher is established outside the EU,
the EU authorized representative is:
Springer Nature Customer Service Center GmbH
Europaplatz 3, 69115 Heidelberg, Germany

Printed by Libri Plureos GmbH
in Hamburg, Germany